Geomorphology and Remote Sensing

Umesh Kumar

AMIGA PRESS. INC
Delhi -110051

AMIGA PRESS INC.
Registered Office:
19 DBS Building, Ansari Road, Darya Ganj, Delhi. 110002, India
Phone: +91-11-23283333 Email: amigapress@gmail.com

This edition is published by Amiga Press.Inc

ISBN : 978-93-84533-11-3

First Published 2018

Reprint 2026

Printed and bound in India by Replika Press Pvt. Ltd.

Preface

Geomorphology is the study of landforms and landscapes, including the description, classification, origin, development, and history of planetar surfaces. During the early part of this century, the study of regional-scale geomorphology was termed "physiography". Unfortunately, physiography also became synonymouswith physical geography, and the concept became embroiled in controversy surrounding the appropriate concerns of that discipline. Some geomorphologists held to a geological basis for physiography and emphasized a concept of physiographic regions, A conflicting trend among geographers was to equate physiography with "pure morphology," divorced of its geological heritage.

In the period following World War II, the emergence of process, climatic, and quantitative studies led to a preference by many Earth scientists for the term "geomorphology" in order to suggest an analytical approach to landscapes rather than a descriptive one. In the second half of the twentieth century, the study of regional-scale geomorphology—the original physiography-was generally neglected. Russell attributed the decline of physiography to its elaborate terminology and to its detachment from evidence acquired by other disciplines, chiefly geology.

Geomorphology, soil science and remote sensing are closely related fields of enquiry through their common interest in the five state-factors of environmental systems: climate, organisms, relief, parent material, and time. Remote sensing, from aerial photography to satellite imagery, constitutes a powerful tool for improving accuracy and precision of extensive large-scale geomorphoiogical surveys, making it possible to investigate previously untestable ideas.

The term 'remote sensing' can generally be defined as "the science and art of obtaining information about an object, area, or phenomenon through the analysis of data acquired by a device not in contact with the object, area, or phenomenon under investigation." This broad definition includes cameras, ocean buoys, and RADAR devices. These general remote sensing devices have undergone adaptations over the last century to make observations of the Earth and its physical process from aircraft and satellites.

Remote sensing technologies can provide the government with the ability to avoid much of the damage caused by unforeseen natural disasters. While weather satellites have monitored hurricanes and tornados since the 1960s, other satellite sensors, such as ETM+ and MODIS, have potential applications for disaster management and response. Scientists have used ETM+ data to monitor patterns in floods, droughts, beach erosion, and volcanic activity over time. MODIS and ASTER data can forecast severe weather with a great degree of reliability, potentially saving states millions of dollars in unnecessary evacuation and emergency response.

This book should be useful to remote sensing scientists and engineers, geographers, geologists, ecologists and environmental scientists, agricultural and soil scientists and urban and regional planners. Regional, state and university remote sensing centres may like to have this book as well.

—*Editor*

Contents

1

Fundamental Concepts of Geomorphology

Most geomorphologists would agree that certain fundamental assumptions underlie all geomorphological investigations. Whether termed "fundamental concepts", "philosophical assumptions", "paradigms", or "basic postulates", these ideas constitute a "conventional wisdom" for the science. One such fundamental concept involves the inherent complexity of landscapes. This concept has impeded the development of grand theories that survive the test of explaining numerous local features.

Another basic assumption involves climatic morphogenesis, emphasizing the role of climatically controlled processes of landform genesis. Several of these concepts have yielded major intellectual controversy, such as the role of cataclysmic processes in shaping the landscape. These concepts apply to geomorphology of all scales.

THE NATURE OF GEOMORPHOLOGY

Geomorphology is the study of landforms and landscapes, including the description, classifiation, origin, development, and history of planetar surfaces. During the early part of this century, the study of regional-scale geomorphology was termed "physiography". Unfortunately, physiography also became synonymouswith physical geography, and the concept became embroiled in controversy surrounding the appropriate concerns of that discipline. Some geomorphologists held to a geological basis for physiography and emphasized a concept of physiographic regions, A conflicting trend among geographers was to equate physiography with "pure morphology," divorced of its geological heritage.

In the period following World War II, the emergence of process, climatic, and quantitative studies led to a preference by many Earth scientists for the term "geomorphology" in order to suggest an analytical approach to landscapes rather than a descriptive one. In the second half of the twentieth century, the study of regional-scale geomorphology—the original physiography-was generally neglected. Russell attributed the decline of physiography to its elaborate terminology and to its detachment from evidence acquired by other disciplines, chiefly geology.

Although the concept of physiographic regions endured among geologists, geographers became much more interested in the details of man/land interactions and in the applications of modeling and systems analysis to geomorphology. In the exploration of planetary surfaces by various space missions, the perspective of regional geomorphology has been the required starting point for scientific inquiry. Global studies of Mars, the Moon, Mercury and Venus resulted in the identification of "surface units" or physiographic provinces.

The Colorado Plateau is an excellent example of a terrestrial physiographic province. Plate I-1 illustrates the use of a large-scale perspective to focus on this naturally defined region. The term "mega-geomorphology" was introduced in March 1981 at the 21st anniversary meeting of the British Geomorpholgy Research Group. The proceedings of that meeting reveal that the concept was not well defined. It clearly involves a return by geomorphologists to the study of phenomena on large spatial scales, ranging from regions to continents to planets.

It also involves large time scales. Nevertheless, mega-geomorphology is merely a convenient term, unencumbered by past philosophical trappings, that emphasizes planetary surface studies at large scales. The interrelation of temporal and spatial scales in geomorphology is illustrated by the tentative classification. Of course, such a hierarchial ordering of geomorphic features is far from satisfying. As stated by Sparks, classifications are arbitrary constructions designed to facilitate the discussionof diverse phenomena at the risk of some distortion of the truth. The scheme merely illustrates what was well known to the great geomorphologists at the last turn of the century. The large first order features, continents and ocean basins, persist and evolve over long time scales. Small high-order features are transient. Fundamental units appear at different orders.

The old concept of physiographic regions was used to designate second-order forms, such as entire mountain ranges or coastal plains. Massive entities within a physiographic region might constitute a third-order form, such as a domal uplift. The details of the classification are unimportant as the analysis moves on to exploring the explanation of phenomena. This book explores mega-geomorphology. The parent science of geology has long emphasized large-scale features in its central discipline of tectonics. Although early proponents of largescale crustal mobilism, such as Alfred Wegener, were decidedly renounced by the mainstream scientific community, their ideas provided the stimulus for work that eventually transformed the Earth sciences.The plate tectonic model that emerged in the late 1960s was but a quantitatively geophysical confirmation of the elegant hypothesis developed by careful attention to large-scale structural patterns on the Earth's surface.

Table: A Hierarchical Classification of Terrestria Geomorphological Features by Scale

Sl. No			Approximate Time Scales of Persistence (years)
1	107	Continents, ocean basins	10^8-10^9
2	10^6	Physiographic provinces, shieds, depositional plains	10^8
3	10^4	Medium-scale tectonic units (sedimentary basins, mountain massifs, domal uplifts)	10^7-10^8
4	10^2	Smaller tectonic units (fault blocks, volcanoes, troughs, sedimentary subbasins, individual mountain zones)	10^7
5	10-10^2	Large-scale erosional/depositional units (deltas, major valleys, piedmonts)	10^6
6	10^{-1}-10	Medium-scale erosional/depositional units or landforms (floodplains, alluvial fans, moraines, smaller valleys and canyons)	10^5-10^6
7	10^{-2}	Small-scale erosional/depositional units or landforms (ridges, terraces, sand dunes)	10^4-10^5
8	10^{-4}	Larger geomorphic process units (hillslopes, sections of stream channels)	10^3
9	10^{-6}	Medium-scale geomorphic process units (pools and riffles, river bars, solution pits)	10^2
10	10^{-8}	Microscale geomorphic process units (fluvial and eolian ripples glacial striations)	-

Of course, this is not intended to imply that microscale studies are unimportant in structural geology. Such studies tell much about the details of rock deformation and the fabric of resulting materials. The session here is that significant science occurs at all scales of study. Scientists neglect the study of one spatial scale to the peril of their advancement to understanding.

SCALES OF STUDY

Callieux and Romani believe that there are two major trends in modern geomorphology: towards quantification and towards more varied extensions. In the latter, they see an extension of concern to other scientific disciplines, to applied problems, to longer time scales, and to more ancient features. They also see an extension to greater spatial scales. The operative temporal and spatial scales of geologic phenomena span an immense range. Note that fluvial phenomena occupy an intermediate position in this scaling.

The various phenomena all follow the general equation:

$$S = aT^b$$

Where S is the size of the feature, T is the time and a and b are constants. The constant b is generally a scaling factor showing that big phenomena tend to last longer. The constant a seems to relate to the intensity of the process. Despite the utility of such scaling relationships, summarized in the concept of allometry problems arise at very large spatial and temporal extrapolations. However, perhaps more important than the details of scaling is the intellectual excitement of multiscale thinking. In his insightful review of geomorphological processes on planetary surfaces, Sharp observed that one of the sessions from the comparative study of landforms on different planets is to "think big."

The same can be said of the application of space-age global remote sensing technology to the study of terrestrial landforms. At thc turn of the century, geomorphologists are once again ready to think big.

Time scales are also important in defining the nature of geomorphic investigations. The role of time scales in geomorphology is discussed more fully by Schumm and Lichty and Cullingford. The macroscale of geomorphic time is that over which major phases of erosion or deposition occur. These phases, which may be cyclic, are controlled by the geological processes of regional warping, mountain building, and crustal plate movement.

The scientific questions arising at the temporal macroscale concern the evolution of the planetary surface over millions of years. Many of the approaches to this time scale are historical, using elements of the landscape to reconstruct past events. Time scales are also important in defining the nature of geomorphic investigations. The role of time scales in geomorphology is discussed more fully by Schumm and Lichty and Cullingford et al.. The macroscale of geomorphic time is that over which major phases of erosion or deposition occur. These phases, which may be cyclic, are controlled by the geological processes of regional warping, mountain building, and crustal plate movement.

The mesoscale of geomorphic time is that which treats major changes in landforms and landscapes over hundreds to thousands of years. Examples include the growth and recession of glaciers, the aggravation and degradation of rivers, and the progradation and recession of shorelines. These changes generally involve a complex interplay between tectonic and climatic controls on geomorphological processes.

This is perhaps the most fertile area of geomorphic analysis, since it involves continual intellectual feedback between historical reconstruction and the study of modern processes.

The microscale of geomorphic time is that over which the major variables of tectonism and climate are assumed to be constant. The processes that characterize sand dunes, glaciers, rivers, or beaches are assumed to reflect only the short-term events that dictate local flow physics.

This is the temporal view of both the engineer and the process geomorphologist. Because of the ability to measure events that occur at the approximate scale of a year, this temporal scale is the richest source of quantitative geomorphic data. The ability to fashion predictions from studies at this scale also makes such studies immensely useful in application. People inhabit the dynamic surface of the Earth and must interact with its surficial processes. Reasoning from one scale to another is an instinctive mental exercise for scientists.

The geomorphologists of the 19th century generally perceived the details of landforms from foot, horseback, or even hands and knees. From such detail, they generalized to the larger synthesis of landscape form and processes. Nevertheless, they also realised the importance of a broad view. The great vistas of the Alps, Appalachians, and Colorado Plateau inspired regional assessment of landscape types.

The reasoning from distant views to the details of site specific investigation is perhaps no better illustrated than in G.K. Gilbert's study of the Henry Mountains in Utah. Gilbert's brilliant concept of the laccolith structure and its modification by erosional processes was conceived before he visited the scene! His field notes show that he conceived the relationships from distant views as he approached the mountains along the Waterpocket Fold.

ORIGIN AND EVOLUTION OF THE EARTH'S CRUST

The Earth's crust is an extremely thin layer of rock, like the skin of an apple in relative terms. It amounts to less than half of 1 per cent of the planet. But the crust is exceptionally important, and not just because we live on it.The crust can be thicker than 80 kilometers in some spots, less than one kilometer in others. Underneath it is the mantle, a layer of rock some 2700 kilometers thick that accounts for the bulk of the Earth. The crust is primarily made of granite and basalt while the mantle beneath is made of peridotite.

Discovery of the Crust

Until just a century ago, we didn't know the Earth has a crust. From astronomical measurements, we knew in the late 1800s that Earth wobbles in relation to the sky as if it had a large, dense core. Beyond that we had no clue, until the advent of seismology. Even today, almost all we know for sure about the deep Earth comes from just one type of evidence: the speed of sound in rock as measured using seismic waves, usually called seismic velocity. The rest is known from intricate, subtle modeling studies. In 1909 the seismologist Andrija Mohorovicic published a paper establishing that about 50 kilometers deep in the Earth there is a sudden change in seismic velocity—a discontinuity of some sort. The discontinuity makes the seismic waves bounce and refract just as light behaves at the discontinuity between water and air. Ever since, that discontinuity, the Mohorovicic discontinuity or "Moho," has been accepted as the boundary between the crust and the mantle beneath.

Crusts and Plates

The crust is not the same thing as the plates of plate tectonics. Plates are thicker than the crust and consist of the crust and the shallow mantle just beneath it; the combination is stiff and brittle and is called the lithosphere. The lithospheric plates lie on top of a layer of softer, more plastic mantle rock that allows the plates to move slowly upon it like a raft in thick mud. We know that the Earth's crust is made of two grand categories of rocks: basaltic and granitic. Basaltic rocks underlie the seafloors and granitic rocks make up the continents. The seismic velocities of these rock types in the lab match the velocities in the crust down to the Moho, so we're pretty sure that the Moho marks a real change in rock chemistry. The Moho isn't a perfect boundary, because some crustal rocks and mantle rocks can masquerade as the other, but even so everyone who talks about the crust, whether in seismological or petrological terms, fortunately means the same thing. In general, then, the crust has two types, oceanic crust and continental crust.

Oceanic Crust

Oceanic crust covers about 60 per cent of the Earth's surface. Oceanic crust is thin and young—no more than about 20 km thick and never older than about 180 million years. Everything older has been pulled underneath the continents by subduction. Oceanic crust is born at the midocean ridges, where pressure upon the underlying mantle is released and the peridotite there begins to melt in response. The part that melts becomes basaltic lava, which rises and erupts while the remaining peridotite becomes depleted.

The midocean ridges migrate over the Earth like Roombas, extracting the basaltic component from the mantle as they go. What that means has to do with rock chemistry. Basaltic rocks contain more silicon and aluminum than the peridotite left behind, which has more iron and magnesium. Basaltic rocks are less dense. In terms of minerals, basalt has more feldspar and amphibole, less olivine and pyroxene, than peridotite. In geologist's shorthand, oceanic crust is mafic while oceanic mantle is ultramafic. Oceanic crust, being so thin, is a very small fraction of the Earth—about 0.1 per cent—but its life cycle serves to refine the rocks of the upper mantle into new rocks with a lighter blend of elements. It also extracts the so-called incompatible elements, which don't fit into mantle minerals and move into the liquid melt. These in turn move into the continental crust as plate tectonics proceeds.

Continental Crust

Continental crust is thick and old—on average about 50 km thick and about 2 billion years old—and it covers about 40 per cent of the planet. Whereas almost all of the oceanic crust is underwater, most of the continental crust is exposed to the air. The continents slowly grow over geologic time as oceanic crust and seafloor sediments are pulled beneath them by subduction. The descending basalts have the water and incompatible elements squeezed out of them, and this material rises to trigger more melting in the so-called subduction factory.

The continental crust is made of granitic rocks, which have even more silicon and aluminum than the basaltic oceanic crust; they also have more oxygen thanks to the atmosphere. Granitic rocks are even less dense than basalt. In terms of minerals, granite has even more feldspar, less amphibole than basalt and almost no pyroxene or olivine, plus it has abundant quartz. In geologist's shorthand, continental crust is felsic.

Continental crust makes up less than 0.4 per cent of the Earth, but it represents the end product of a double refining process, first at midocean ridges and second at subduction zones. The total amount of continental crust is slowly growing. The incompatible elements that end up in the continents are important because they include the major radioactive elements uranium, thorium and potassium.

They create heat, which makes the continents act like electric blankets on top of the mantle. The heat also softens thick places in the crust, like the Tibetan Plateau, and makes them spread sideways. Continental crust is too buoyant to return to the mantle. When continents collide, the crust can thicken to almost 100 km, but that is temporary. The limestones and other sedimentary rocks that form on the continents are likewise lighter than basalt. Even the sand and clay that is washed off into the sea returns to the continents on the conveyor belt of the oceanic crust. Continents are truly permanent, self-sustaining features of the Earth's surface.

THE GEOLOGIC COLUMN

The *standard geologic column* is the basis for the geologic time scale. The standard column is composed of a sequence of European formations because it was developed by expanding the stratigraphic sequences worked out by Smith, Cuvier, and Brongniart.

Its fossil assemblages form the standard of comparison for sequences in all parts of the world. Correlations from one part of a continent to a distant part, or even to another continent, are based on assemblages of fossil plants and animals. Such correlations are sometimes handicapped by the local, rather than cosmopolitan, distribution of many fossils and by differing conditions of deposition. On the other hand, they are facilitated by the vast wealth of former living things; thousands upon thousands of varied forms are preserved in the fossil record. The *standard time scale* is a scale representing the periods of deposition of the rock units in the standard stratigraphic column.

Simple Early Column

As early as the middle of the Eighteenth Century, Italian and German geologists had classed their local rocks into three groups: Primary (rocks like granite and gneiss, with neither bedding nor fossils), Secondary (cemented sedimentary rocks, generally exposed in the mountains), and Tertiary (weakly consolidated sedimentary rocks of the lowlands, which rested upon the Secondary rocks). Though fossils were known from both Secondary and Tertiary rocks, distinctions among them from bed to bed were either not made or not used in the Eighteenth Century classification.

Standard Column

The present standard column was pieced together in the Nineteenth Century from strata exposed in various parts of Europe. The major subdivisions of the column are called *rock systems*. Most of the systems are represented, at least in part, in the area of Britain mapped by Smith and his early followers.

Smith grouped the oldest strata in his section as the *Old Red Sandstone*. These rocks contain few fossils and, in fact, are largely land-laid beds, but to the south they interfinger with fossiliferous marine beds widely represented in Devonshire.

he Old Red Sandstone is, therefore, now considered a part of the Devonian System. Resting on the Old Red Sandstone is a group of strata called by Smith the *Mountain Limestone,* and this series is in turn overlain by a succession of sandstones, shales, iron ores, and coals which Smith called the *Coal Measures*. These rocks and the underlying Mountain Limestone are now grouped together as the *Carboniferous System,* named from the abundance of coal in the rocks.

Above the Coal Measures Smith recognized a series of beds called the *Magnesian Limestone,* followed in turn by the *New Red Sandstone.* Both these formations are poorly fossiliferous and largely of continental rather than marine origin. The Magnesian Limestone is now considered part of the *Permian System,* named from a province in Russia just west of the Ural Mountains, where marine fossiliferous rocks occur in the same stratigraphic position. The New Red Sandstone, on similar grounds, is now regarded as part of the *Triassic System,* named from the sandwich of a marine limestone between two continental formations found just above the Permian System in Germany.

Overlying the New Red Sandstone, Smith recognized a whole series of richly fossiliferous beds which he divided into many formations, partly because the abundant fossils facilitated correlations and partly because the canals he constructed lay across these beds, and he thus had ample time to work out the details of their successions. Some years later many identical fossils were found in the rocks of the Jura Mountains of Switzerland and France. The name *Jurassic System,* to which these strata are now referred, recalls this fact. Jurassic rocks also underlie the Lower Chalk of the Paris Basin.

Toward London, the Jurassic strata are overlain by greensand, and this in turn by chalk. This series is grouped as the *Cretaceous System* (named from the Greek word for chalk). The fossils are like those of the chalk in the Paris Basin. Overlying the chalk, Smith found the *London Clay,* a group of beds rich in fossils that are like the fossils in the beds above the chalk of the Paris Basin. More than a generation after Smith, the stratigraphy of the highly disturbed rocks below his Old Red Sandstone was worked out. Three new systems were recognized: in descending order, *Silurian, Ordovician* (both names of old British tribes), and *Cambrian,* from the Latin name for Wales.

The Cambrian strata are the oldest rocks that contain fossils in any abundance, although sedimentary rocks thousands of feet thick are found beneath them. In this book such older rocks are grouped as the *pre-Cambrian.* They have great bulk and complexity and can be classified in various ways, in spite of the absence of fossils. Perhaps refinements in the dating of rocks by radioactivity may some day enable us to classify the pre-Cambrian rocks in an orderly chronological sequence, just as fossils have permitted us to do for the Cambrian and younger systems.

Geologic Time Scale

The standard geologic column is the basis for the *geologic time scale.* The names applied to the systems of rocks are also used for the *periods of time* during which the respective systems of rocks were deposited. Thus, we use the term *Carboniferous Period* for the time during which the *Carboniferous System*—the Mountain Limestone and Coal Measures of Smith—were laid down. Appended to it are remarks as to the sources of the names and the ages of several minerals as found from studies of radioactivity. From these ages in years it is apparent that more subdivisions are recognized among younger rocks than among older. Just as in human history, the documents are more numerous and the gaps in knowledge less wide the more recent the period with which we deal.

Gaps in the Time Scale

Divisions of the standard column are based on abrupt changes in the fossil assemblages of the strata in Europe. These points of division naturally were those marking the longer intervals of erosion or nondeposition in the European stratigraphic series. The longer the interval lost by erosion, the more prominent the fossil differences would be. As stratigraphic work was extended to other continents, however, fossil assemblages intermediate between those of adjacent periods as recognized in western Europe were discovered.

With more and more stratigraphic work the gaps in the type sections have continued to decrease. Beds containing fossil assemblages that pose "boundary problems"—that is, uncertainty as to their correlation with the upper part of one European system or the lower part of the next higher—are thus present at one place or another for nearly every systemic boundary.

Even the boundary between the Paleozoic and Mesozoic Eras, long thought to represent an interval during which none of the present continents received deposits, is apparently bridged by an essentially complete succession of beds in the Himalayan Mountains. Even in Europe, where the divisions were first established, further research has served to narrow the lost gaps in the record. For example, a new division, the Paleocene, has been recently established for strata intermediate in age between Eocene and Cretaceous, where a long gap had been thought to exist.

Now we find that different paleontologists refer the same strata (let us say, in Wyoming) to the Cretaceous or to the Paleocene, depending on their respective appraisals of the relative affinities of the fossils of these beds to that of one or another of the European sections.

Intercontinental Correlation

Thus, although the original divisions of the type section were based on stratigraphic relations of the beds, our correlations of distant strata are necessarily based on a *comparison of fossils*, not strata. The demonstration of an interruption in stratigraphic succession in the type area can give no help in correlating beds in a distant part of the world. A stratigraphic break in Nevada, for example, cannot be used as evidence that the beds above it are Permian and those below Carboniferous. This assignment of age to the Nevada strata can be made only on the basis of comparison of fossils with those of the type section of Europe. The uncertainties posed by "boundary problems," however, are concerned largely with form instead of fact. Such disputed strata, it is generally agreed, must represent times between, or at least partially bridging, the gaps in the type areas of Europe. Further detailed stratigraphic work cannot but fill other parts of the gaps. This is indeed a tribute to the precision of correlation by fossils rather than a criticism of its uncertainties.

The Duration of Geologic Time

With the building of the geologic time scale there came a growing appreciation of the vast duration of geologic time. Contemplation of even the time required to lay down the 500-foot thickness of chalk in the Paris Basin is upsetting to the man who thinks of time only in terms of the span of human life. Some strata of the chalk are composed of the skeletons of minute animals and plants. Similar deposits are accumulating today at rates so low as to defy measurement—certainly no more than a few inches per century and probably much less. Yet the chalk represents only a small part of Cretaceous time, and the Cretaceous itself is a mere fraction of geologic time.

Not only the vast thickness of sediment impressed geologists with the immensity of geologic time, but also the vast parade of life recorded by the fossils—the development of thousands upon thousands of new species with advancing time and the gradual dying out of whole fossil assemblages and their succession by others—time after time—can only be conceived of as involving millions of years, unless greatly different rates of change prevailed in the past.

Time Represented by the Geologic Column

Spurred by their curiosity to translate geologic time into years, geologists tried many approaches. The simplest and most obvious one was to add the greatest thickness of strata of each period and thereby get the total thickness for all geologic time. This total, divided by the present annual rate of sedimentation, might be thought to give the length of geologic time in years.

However, so many assumptions are involved that such a calculation becomes meaningless. How can we arrive at an average annual rate of sedimentation when we know that it may take a century to lay down less than an inch of chalk, whereas a desert cloudburst may deposit 40 feet of gravel, sand, and boulders overnight?

And how can we measure the time of scour and nondeposition in marine sediments when we know that a single storm may remove the accumulation of years from part of the sea bottom and perhaps pile up more sand on a nearby beach than was deposited during the whole preceding decade? In short, the rate of sedimentation varies so greatly and has been so little measured that the average annual rate can only be guessed at. This method can only yield an order of magnitude for geologic time. By such a calculation, the British geologist Sollas estimated (1899) that 34 to 75 million years have passed since the beginning of the Paleozoic —the larger figure including a guess at the length of the lost intervals in the record. These figures have a certain interest but surely are not reliable.

Age of the Ocean

The Irish scientist Joly made a more ingenious approach. He reasoned that the salt in the sea must have been brought in by the rivers, after being weathered from the rocks. Most of it has accumulated in the sea, as little is blown out by the wind or deposited by evaporation of sea water in desert lagoons or on tide flats.

Hence the oceans must be growing saltier; and, if the amount of salt they now contain could be measured and divided by the annual increment, their age might be obtained. From the many elements in solution in the oceans, Joly selected sodium as the constituent best suited to this method.

The volume of the sea has been roughly computed from its average depth and area, and its composition is known from thousands of analyses. Thus the total sodium it contains can be roughly computed. Similarly, we have thousands of analyses of river waters, and from stream gages we know fairly well how much river water flows annually to the sea. From these figures Joly obtained a rough estimate of the annual increment of sodium. Dividing the first amount by the second, he found:

Joly realized, of course, that this computation disregarded many factors. For one thing, river discharge was probably not constant throughout geologic time. Furthermore, not all the sodium in rivers is derived from weathering; much is from sewage and industrial waste, some is wind-blown from sea beaches, and much is leached from former marine sediments. Thus the present annual increment is higher than in the geologic past. How much higher cannot be ascertained.

Again, there are great deposits of rock salt among the stratified rocks, derived from evaporated sea water. If returned to the sea, they would increase the tonnage in the oceans. Again we have no way of getting an accurate correction. We have noticed how magnesium ion reacts with limy ooze to form dolomite. In a somewhat similar way, we now know that sodium ion reacts with clay minerals and is incorporated into marine muds and thus removed from solution. Finally, large amounts of sea water are mechanically enclosed in the pores of marine sedimentary rocks and the sodium so contained is then removed from the ocean. When all these corrections are considered, it is obvious that Joly's 99 million years must be multiplied many fold, but, as no figure can be made quantitative (that is, subject to careful measurement for purposes of verification), the results yield nothing but the fact that the ocean has existed for a vast span of time.

Age Based on Classical Physics

Another estimate of the duration of geologic time was made by Lord Kelvin, the famous English physicist of the late Nineteenth Century. In his day, before the discovery of radioactivity, atom-splitting and reconstitution, and transformation of mass to energy, Kelvin could imagine no source of energy for the stars except the contraction of the gases composing them.

Every astronomical body, including the earth, radiates heat into outer space, he knew, and, therefore, decided that each must be cooling and hence shrinking in size. He computed that, even if the earth had contracted from a size so great that it originally touched the sun, no reasonable assumptions as to original temperature could allow its age to exceed 400 million years. In fact, it was probably only about 20 million years old.

Lord Kelvin's dictum—despite his prestige as the leading physicist of his day—was rejected by geologists. Though the rates of sedimentation and of organic evolution cannot be measured, no geologist could bring himself to accept so short a time as Kelvin advocated. They regarded it as "unreasonably" short, despite his exact mathematical demonstration. We now know that his basic assumptions were the source of his error. The discovery of radioactivity and of other sources of energy not known to Kelvin demolished his basic assumption that all stellar energy arises from contraction. The discovery of radioactivity also made it possible to establish some points on the geologic time scale in terms of years, and to show that all previous estimates of the duration of geologic time were far too short.

The Radioactive Clock

When Becquerel discovered radioactivity (1898), he opened new vistas in every science. One of the many byproducts of this discovery was the determination of some geologic ages in years.

A few elements, among them uranium and thorium, disintegrate spontaneously into other elements. The atomic nuclei of such elements are intrinsically unstable and emit *alpha* particles and *beta* particles (electrons). Each nuclear emission, whether of an alpha or beta particle, transforms the atom into a different element. Starting with uranium 238 (U_{238}) there are 15 steps in this natural process: 8 alpha particles and 7 beta particles are emitted, yielding ultimately a stable form of lead (Pb_{206}) which does not disintegrate further.

The rate of spontaneous disintegration varies tremendously with different elements. It is expressed in terms of the element's "half-life," which is the time required for half its atoms to disintegrate. The half-life of some members of the U_{238} series is only a fraction of a second; for U_{238} itself it is millions of years. The half-life period of the entire chain from U_{238} to Pb_{206} has been determined to be 7,600 million years. In other words, if we start with 1 gram of U_{238}, in 7,600 million years only 0.5 gram will be left, in another 7,600 million years there will be only 0.25 gram, and so on.

The other half (or three-fourths) has changed into lead, helium ions, electrons, and very small amounts of intermediate elements in the series. In no experiment has the disintegration rate been changed by heat, pressure, state of chemical combination of the element, or time. The half-life of a radioactive element is thus considered a constant, that is, a fundamental property of the element.

Many minerals, most comparatively rare, contain appreciable amounts of uranium. Some of these are intimately associated with intrusive igneous rocks and have been segregated from the magma during the last stages of its consolidation by their occurrence in dikes and pods that grade into the intrusive rock. By analyzing such minerals and finding the ratio of uranium-derived lead to the uranium still remaining it is possible to find their age, and hence the time in years since the magma from which they crystallized was consolidated.

Such analyses are difficult. The proportions of the different lead isotopes must be determined so that uranium-derived lead (Pb_{206})can be differentiated from any ordinary lead that may be present. Ordinary lead is a mixture of isotopes with an atomic weight of 207.21. Furthermore, all uraniumbearing minerals contain U_{238} as well as the more abundant U_{238}, and may contain thorium. Each of these also disintegrates to yield a lead having a different atomic weight.

These additional radioactive substances make it difficult to find the ratios accurately. Absolutely fresh unweathered minerals must be used, for solutions circulating long after the mineral was formed might leach out the lead and uranium at different rates and thus produce great errors in the calculation of the age. Though many determinations have been made, only a few meet the rigid requirements with regard to analytical procedure and freshness of material.

Thus we have a fairly reliable method for dating geologic time in years. Furthermore, for the rocks older than the Cambrian, it is the only method available for interregional correlation. The oldest mineral from the United States (1,600 million years) thus far determined comes from the Black Hills of South Dakota. It was taken from a segregation in a granite mass that cuts across schists.

Therefore, it must be younger than the schists, which for many years had been considered late pre-Cambrian because they have not been severely metamorphosed. The Grenville gneiss of the Adirondacks and southern Canada, on the other hand, is highly contorted and severely metamorphosed, so that it has been regarded as extremely old pre-Cambrian. It turns out from the lead ratios that this rock is little more than one-third as old as the Black Hills rock. Such evidence shows the impossibility of using the degree of deformation and metamorphism of the rocks as a basis of correlation over long distances.

CONTEMPORARY GEOMORPHOLOGY

Today, the field of geomorphology encompasses a very wide range of different approaches and interests. Modern researchers aim to draw out quantitative "laws" that govern Earth surface processes, but equally, recognize the uniqueness of each landscape and environment in which these processes operate. Particularly important realizations in contemporary geomorphology include:

1) that not all landscapes can be considered as either "stable" or "perturbed", where this perturbed state is a temporary displacement away from some ideal target form. Instead, dynamic changes of the landscape are now seen as an essential part of their nature.
2) that many geomorphic systems are best understood in terms of the stochasticity of the processes occurring in them, that is, the probability distributions of event magnitudes and return times. This in turn has indicated the importance of chaotic determinism to landscapes, and that landscape properties are best considered statistically. The same processes in the same landscapes does not always lead to the same end results.

Processes

Modern geomorphology focuses on the quantitative analysis of interconnected processes. Modern advances in geochronology, in particular cosmogenic radionuclide dating, optically stimulated luminescence dating and low-temperature thermochronology have enabled us for the first time to measure the rates at which geomorphic processes occur. At the same time, the use of more precise physical measurement techniques, including differential GPS, remotely sensed digital terrain models and laser scanning techniques, have allowed quantification and study of these processes as they happen. Computer simulation and modelling may then be used to test our understanding of how these processes work together and through time.

Geomorphically relevant processes generally fall into (1) the production of regolith by weathering and erosion, the transport of that material, and its eventual deposition. Although there is a general movement of material from uplands to lowlands, erosion, transport, and deposition often occur in closely-spaced tandem all across the landscape.

The nature of the processes investigated by geomorphologists is strongly dependent on the landscape or landform under investigation and the time and length scales of interest. However, the following non-exhaustive list provides a flavour of the landscape elements associated with some of these.

Primary surface processes responsible for most topographic features include wind, waves, chemical dissolution, mass wasting, groundwater movement, surface water flow, glacial action, tectonism, and volcanism. Other more exotic geomorphic processes might include periglacial (freeze-thaw) processes, salt-mediated action, or extraterrestrial impact.

Fluvial Processes

Rivers and streams are not only conduits of water, but also of sediment. The water, as it flows over the channel bed, is able to mobilize sediment and transport it downstream, either as bed load, suspended load or dissolved load. The rate of sediment transport depends on the availability of sediment itself and on the river's discharge. Rivers are also capable of eroding into rock and creating new sediment, both from their own beds and also by coupling to the surrounding hillslopes. In this way, rivers are thought of as setting the base level for large scale landscape evolution in nonglacial environments. Rivers are key links in the connectivity of different landscape elements.

As rivers flow across the landscape, they generally increase in size, merging with other rivers. The network of rivers thus formed is a drainage system and is often dendritic, but may adopt other patterns depending on the regional topography and underlying geology.

Aeolian Processes

Aeolian processes pertain to the activity of the winds and more specifically, to the winds' ability to shape the surface of the Earth. Winds may erode, transport, and deposit materials, and are effective agents in regions with sparse vegetation and a large supply of unconsolidated sediments. Although water and mass flow tend to mobilize more material than wind in most environments, aeolian processes are important in arid environments such as deserts.

Hillslope Processes

Soil, regolith, and rock move downslope under the force of gravity via creep, slides, flows, topples, and falls. Such mass wasting occurs on both terrestrial and submarine slopes, and has been observed on Earth, Mars, Venus, Titan and Iapetus. Ongoing hillslope processes can change the topology of the hillslope surface, which in turn can change the rates of those processes.

Hillslopes that steepen up to certain critical thresholds are capable of shedding extremely large volumes of material very quickly, making hillslope processes an extremely important element of landscapes in tectonically active areas.

On Earth, biological processes such as burrowing or tree throw may play important roles in setting the rates of some hillslope processes.

Glacial Processes

Glaciers, while geographically restricted, are effective agents of landscape change. The gradual movement of ice down a valley causes abrasion and plucking of the underlying rock. Abrasion produces fine sediment, termed glacial flour. The debris transported by the glacier, when the glacier recedes, is termed a moraine. Glacial erosion is responsible for U-shaped valleys, as opposed to the V-shaped valleys of fluvial origin.

The way glacial processes interact with other landscape elements, particularly hillslope and fluvial processes, is an important aspect of Plio-Pleistocene landscape evolution and its sedimentary record in many high mountain environments. Environments that have been relatively recently glaciated but are no longer may still show elevated landscape change rates compared to those that have never been glaciated. Nonglacial geomorphic processes which nevertheless have been conditioned by past glaciation are termed paraglacial processes. This concept contrasts with periglacial processes, which are directly driven by formation or melting of ice or frost.

Tectonic Processes

Tectonic effects on geomorphology can range from scales of millions of years to minutes or less. The effects of tectonics on landscape are heavily dependent on the nature of the underlying bedrock fabric that more less controls what kind of local morphology tectonics can shape. Earthquakes can, in terms of minutes, submerge large extensions creating new wetlands.

Isostatic rebound can account for significant changes over thousand or hundreds of years, and allows erosion of a mountain belt to promote further erosion as mass is removed from the chain and the belt uplifts. Long-term plate tectonic dynamics give rise to orogenic belts, large mountain chains with typical lifetimes of many tens of millions of years, which form focal points for high rates of fluvial and hillslope processes and thus long-term sediment production. Features of deeper mantle dynamics such as plumes and delamination of the lower lithosphere have also been hypothesised to play important roles in the long term (> million year), large scale (thousands of km) evolution of the Earth's topography. Both can promote surface uplift through isostasy as hotter, less dense, mantle rocks displace cooler, denser, mantle rocks at depth in the Earth.

Igneous Processes

Both volcanic (eruptive) and plutonic (intrusive) igneous processes can have important impacts on geomorphology.

The action of volcanoes tends to rejuvenize landscapes, covering the old land surface with lava and tephra, releasing pyroclastic material and forcing rivers through new paths.

The cones built by eruptions also build substantial new topography, which can be acted upon by other surface processes.

Biological Processes

The interaction of living organisms with landforms, or biogeomorphologic processes, can be of many different forms, and is probably of profound importance for the terrestrial geomorphic system as a whole.

Biology can influence very many geomorphic processes, ranging from biogeochemical processes controlling chemical weathering, to the influence of mechanical processes like burrowing and tree throw on soil development, to even controlling global erosion rates through modulation of climate through carbon dioxide balance.

Terrestrial landscapes in which the role of biology in mediating surface processes can be definitively excluded are extremely rare, but may hold important information for understanding the geomorphology of other planets, such as Mars.

Scales in Geomorphology

Different geomorphological processes dominate at different spatial and temporal scales. Moreover, scales on which processes occur may determine the reactivity or otherwise of landscapes to changes in driving forces such as climate or tectonics. These ideas are key to the study of geomorphology today.

To help categorize landscape scales some geomorphologists might use the following taxonomy:

- 1st-Continent, ocean basin, climatic zone (~10,000,000 km^2)
- 2nd-Shield, e.g. Baltic Shield, or mountain range (~1,000,000 km^2)
- 3rd-Isolated sea, Sahel (~100,000 km^2)
- 4th-Massif, e.g. Massif Central or Group of related landforms, e.g., Weald (~10,000 km^2)
- 5th-River valley, Cotswolds (~1,000 km^2)
- 6th-Individual mountain or volcano, small valleys (~100 km^2)
- 7th-Hillslopes, stream channels, estuary (~10 km^2)
- 8th-gully, barchannel (~1 km^2)
- 9th-Meter-sized features.

Overlap with other Fields

There is a considerable overlap between geomorphology and other fields. Deposition of material is extremely important in sedimentology. Weathering is the chemical and physical disruption of earth materials in place on exposure to atmospheric or near surface agents, and is typically studied by soil scientists and environmental chemists, but is an essential component of geomorphology because it is what provides the material that can be moved in the first place. Civil and environmental engineers are concerned with erosion and sediment transport, especially related to canals, slope stability (and natural hazards), water quality, coastal environmental management, transport of contaminants, and stream restoration. Glaciers can cause extensive erosion and deposition in a short period of time, making them extremely important entities in the high latitudes and meaning that they set the conditions in the headwaters of mountain-born streams; glaciology therefore is important in geomorphology.

GEOGRAPHERS AND THE THEORY

Fieldwork, especially in "foreign" lands, is one of the time-honoured hallmarks of professional geography, yet as Felix Driver and others have noted, "little attention has been paid to the specifically geographical dimensions of field-work [sic] or to its history within the discipline." In this chapter I will explore pieces of this history through the practice of return visitation to particular places, with an emphasis on the record of North American geographers' engagement with Latin America. In turn, Carl Sauer's Mexican research, together with that of some of his students and others inspired by his example, appears to be the most tightly woven net of repeat visitation. Here, one finds multiple strands of crisscrossing lines of reconnaissance and research, and complicated proxy relations involving individuals from later generations pursuing topics initiated by individuals from earlier ones.

Space here does not permit more than sketching the outlines of this history, but it does illustrate some of the salient aspects of repeat visitation, and points to a neglected topic in the history of geography. Viewed from the broad perspective of geography's historical record, the practice of return or repeat visits appears to be a more recent development than a time-honoured tradition. Various factors can be proposed as contributing to this apparent trend towards repetition. Perhaps the foremost is geography's shift from a spatial-descriptive exploratory enterprise to a modern scientific discipline with a premium placed on deriving generalisations through replicable results. Although few of the human geographers in the past century engaging in foreign area research have been self-conscious devotees of approaches involving nomothetic, spatial analytical science, the ascendancy of a "scientific" human geography in the post-WWII period can be implicated in the decline of chorology and geographic description for its own sake.

A second impetus also might be seen as more influence from outside geography than from within. Along with the rise of a more analytic, "scientific" human geography, geographers began to interact more with other social scientists, and borrow from their theory and practice. For field-oriented geographers working abroad, especially cultural geographers, the paths most often crossed or followed were anthropological. Increasingly archaeologists and ethnologists were digging in and making long investments in discrete study sites, rather than mounting wide-ranging surveys and broad collecting ventures. A variant of this development, especially for geographers, is repeat visit research at intervals to measure change (or lack of it) in places, landscapes, or regions.

Elements of the mobility and transience of the older survey and collection mode is carried forward, but with an emphasis on demonstrating familiarity rather than novelty. A fourth influence that began to emerge by the 1960s and was accelerated by that decade's ferment, involved ethical questions and issues. Anthropologists, followed by other social scientists including geographers, generally came to see and accept that not only their research had varying impacts on their "subjects," but also that forms of reciprocity might be part of a researcher's protocol and practice. Among other modes, this has taken the form of extended ties to individuals or communities, often spanning decades since the initial research was conducted (Richardson 1998). In varying degrees, all of these underlying factors may play a part in the decisions and the designs that have led geographers to return at intervals, often multiple times, to favoured sites, landscapes, or regions. I suspect, however, that most returnees would simply say that the main rationale or attraction is that they have come to identify strongly with their adopted site or region and feel a need or obligation to return.

In some cases it may serve as a second home, and contrary to the old saw, familiarity here breeds not contempt, but affection and attachment. James Parsons, perhaps better than anyone, evocatively makes the case for this in his AAG presidential address: "Repeated visits and growing familiarity with an area continually enlarge one's range of vision and concern. You see, you hear, or read something new, something out of place, something irregular, and you are moved to investigate further." His own periodic return visits to Antioquia in Colombia reverberated in his lifetime and continue to resonate there today. As with much of modern geography, the groundings and ground floor for both the theory and practice of foreign area fieldwork can be found in German scholarship.

By the late 19th century geographic training in German universities followed a regimen involving lecture courses, research seminars, and perhaps paramount, fieldwork. Typically a graduate student was first expected to conduct fieldwork and produce a thesis on a topic of local focus, preferably in one's Heimat or "home" locale or region. This exercise was part of the larger Heimatkunde or local study movement that was integral to the formation of German nationalism. For students completing doctoral study, a dissertation based on foreign area fieldwork was expected. In turn, foreign area studies, or Auslandskunde became an integral component of German colonial exploration and expansion as well as demonstrating the prowess of German science and scholarship to a world audience. The exemplars of the Heimatkundel Auslandskunde tradition in German geography before it became formalised in university training, were Johann Gottfried Herder and Alexander von Humboldt. Herder valorised local studies, especially of folk culture, and coined the term "nationalism" and "populism" among others.

Humboldt provided a towering example of foreign field studies with his five years of travel and investigations in South America, Mexico, and Cuba, and later in Russia and Central Asia (Livingstone 1992, Matless 1992). Both Herder and Humboldt were complex Enlightenment figures that also gave birth and voice to major tenets of Romanticism. It could be argued that their combinations of the Enlightenment penchant for empirical scientific discovery with the Romantic impulse for direct experience, especially of "nature," provided a template for future fieldwork practice in geography. One example that Humboldt did not provide, was of long-term engagement with any particular place or region. Although he could be an acute observer of local phenomena, his sights were ultimately set on framing and answering global scale questions and problems. For this, mobility and breadth of experience, not in-depth local immersion, was called for.

Moreover, he was the consummate cosmopolitan, and was content to process the wealth of empirical data that flooded in to him by his global network of colleagues from his bases in Paris and later Berlin. Herder, on the other hand, in valorising local studies, made immersion in local circumstances, including learning local languages and customs a badge of distinction. By the turn of the twentieth century, emergent national traditions in geographic practices and perspectives were beginning to coalesce in university contexts beyond Germany. The "home geography" followed by a foreign area study model may have been an ideal, but it was far from obligatory. French geographers found the geographical "other" in their own regional distinctiveness. Russian geographers could travel continental distances but still not be abroad (Bassin 1999). British geographers had the regions and resources of a global empire to construct and catalogue, but much of this could be done at home, and was. In the early years of professional geography, North American geographers largely confined their scope and ventures to national space.

There were, of course, exceptions. As early as the 1870s William Morris Davis spent three years in Cordoba, Argentina conducting research at the Argentine Meteorological Observatory. A decade later Mark Jefferson (1884-1889) also spent time at the Cordoba observatory and worked as a submanager of a sugar plantation near Tucuman before beginning his studies with Davis at Harvard (Martin 1968). Wellington Jones, co-author with Carl Sauer of an early and key statement on fieldwork (1915), assisted Bailey Willis with physiographic survey for rail lines in northern Patagonia during 1911-1912. Although these were among the first Latin American fieldwork efforts by North American geographers, none of the three made Latin America the main focus of their research and only Jefferson returned to do research.

This was in 1918 under the auspices of the American Geographical Society's ABC Expedition. Isaiah Bowman, the AGS director, commissioned Jefferson and an assistant to study European immigration in the Southern Cone, especially Germans, in the context of WWI and its aftermath. Throughout the 1920s Jefferson continued to publish the results of the several months he spent in Chile, Argentina, and Brazil in 1918, but he did not make them a return destination. Nor did he direct graduate students to the Southern Cone, as he spent his career teaching at Ypsilanti Normal School. Similarly Isaiah Bowman accumulated a wealth of experience and knowledge with his three field seasons in Andean Peru, Bolivia, and Chile.

These trips resulted in a number of important publications (1916, 1924) but like Jefferson, South America was not a repeat destination, and Bowman's role as AGS Director (1915-1935) and President of Johns Hopkins University (1935-1949) precluded overseeing student research. Although as AGS Director he did foster and direct resources to Latin American geographic research (Martin 1980). It wasn't until academic geography's second generation in the US, or with those that began their professional careers in the post-WWI period, that sustained engagement in foreign area fieldwork took hold. For Latin America, two nodes developed coalescing around two distinctly different figures—Preston E. James and Carl O. Sauer. James came from a prosperous New England family wherein a Harvard education was not an option—it was an expectation. At Harvard he discovered geography and completed his first two degrees in geography, continuing on with doctoral studies in geography at Clark (Martin 1987). In contrast, Sauer came from a Midwestern German immigrant background, and found geography via geology in graduate school at Chicago.

Though ten years Sauer's junior, James' first forays into Latin America preceded Sauer's by several years. At the end of his first term at Clark (April 1921) he set off for an extended tour of South America. He took a United Fruit boat to Colombia, then down the Andes to Chile, through Argentina, Uruguay and Brazil, and back by boat via the Lesser Antilles. From this trip he gathered enough data and observations to publish a half dozen articles. It also gave him his dissertation topic—transportation in South America. In 1923 he replaced Sauer at Michigan, and his next field trip was to Trinidad in 1924. This resulted in another six publications. In 1930, 1938 and in 1949 as a visiting lecturer he did extensive fieldwork in Brazil. These two decades of engagement with Brazil yielded some two-dozen publications.

After WWII James travelled to a number of places in Latin America in concert with his official duties with the Pan American Institute of Geography and History (PAIGH) and in other capacities, augmenting his compendious textbook on Latin America, but like Isaiah Bowman, once he assumed a heavy load of responsibilities beyond teaching and research, fieldwork was essentially shelved, and he doesn't seem to have been concerned to return to sites once studied or observed. James' approach was broad and comprehensive, publishing on a wide range of topics from climatology and geomorphology to transportation and economic development to historical and cultural landscapes, in a number of countries. Sauer's trajectory was different. Once he left Michigan and the Midwest for Berkeley in 1923, he also developed a life-long engagement with Latin America, but only in select places and via a narrower range of geographical topics—mostly cultural and historical. Sauer's introduction to Latin America came through his musing that Baja California was "the mother of California historically" and should be studied as such.

This notion sparked interest in some of his graduate students, and Peveril Meigs and Warren Thornthwaite made a reconnaissance of the northern portion of the peninsula in the summer of 1925. Their seminar reports that fall convinced Sauer that his future lay south of the border, especially since he felt California had already been staked out and claimed by other researchers, particularly geologists, historians, and anthropologists. His first venture into Mexico was during the summer of 1926, or five years after James' first South America circuit. He went with Meigs and Fred Kniffen. They retraced Meig's itinerary the previous summer, travelling south as far as San Fernando, or about a quarter of the way down the peninsula. Sauer was primarily interested in studying the geomorphology, but Meig's interest in the Spanish missions helped turn Sauer towards the historical features in the landscape as well. Both Meigs (1932) and Kniffen (1930) went on to write dissertations on northwest Mexican topics, but neither made Latin America "their region." Sauer planned a follow-up field season in 1927 to check on Meigs and Kniffen in the field, but other things intervened.

The summer of 1928 he moved his sights east, devoting seven weeks to exploring northeastern Sonora. Plans to spend the summer of 1929 in Sonora were thwarted by political unrest there, but Sauer received a sabbatical leave for the spring of 1930 and he made the most of it. He left in late December 1929 with his family, and post-doe Gottfried Pfeifer for Mazatlan, Sinaloa. They were joined shortly by student Donald Brand and at various times during the season, anthropologists Alfred Kroeber and Paul Kirchoff and artist Antonio Sotomayor. In Sinaloa they found archaeological evidence of advanced pre-Hispanic cultures, thus Sauer was able to propose new boundaries for Mesoamerica's northwestern extension (Sauer and Brand 1932). In late March they moved their field headquarters to Nogales, Arizona. From this base they mounted "archaeogeographical" surveys of northwestern Sonora, discovering extensive trinchera (habitational terraces) settlement sites (Sauer and Brand 1931).

Sauer and Brand spent another three months in the field and also the regional archives. Winter break 1930-1931 Sauer was back in Sonora, this time accompanied by Joseph Spencer, a new graduate student from UCLA. Spencer stands out as one of Sauer's students who very much charted his own course, not following Sauer's Latin American path, but making East and Southeast Asia his career choice for fieldwork and return visits.

They spent three weeks, mostly on horseback, in central Sonora on reconnaissance for trinchera sites. In early 1931, based on his previous northwest Mexican fieldwork and discoveries, Sauer was awarded a prestigious John Simon Guggenheim Memorial Fellowship to continue this work. Not only did this put Sauer back into the field for fall semester 1931, but it also led to Sauer's wider participation in the Guggenheim Fellowship programme as a member of the selection committee from 1934 to the mid-1960s. Sauer's recruitment into the Guggenheim ranks led to associations with other East Coast establishment foundations. He forged close ties with Henry Allen Moe, the Guggenheim program's secretary.

This led to introductions and involvements with the Rockefeller Foundation's Division of Social Sciences and the Carnegie Foundation, and in turn, support for Sauer and his students' Latin Americanist fieldwork. Sauer's "Guggenheim Year" (summer and fall of 1931) in Mexico expanded his experience of both field and archive to Central Mexico In the process, he became increasingly familiar with trails in the Sonora and Sinaloa. This time his graduate assistant was Leslie Hewes, who like Spencer, did not elect Latin America for career commitment, choosing instead a Heimat theme—the Cherokee country in his own native Oklahoma for his dissertation and then the wider Great Plains for his subsequent work (Wishart 2000). Sauer was not successful in finding the sought-after evidence for demonstrating links between the high culture areas of western Mexico and the US Southwest, one of the objectives of his Guggenheim fellowship. He did, however, gather material for his acclaimed study, "The road to Cibola" (1934b), an interpretation of sixteenth-century Spanish exploration and travel along the west coast of Mexico northward to the US Southwest.

This publication, more than his previous efforts, brought Sauer to the attention of historians and others with interests in colonial exploration. His next venture took him to Chihuahua and Durango in the summer of 1933. Along with a graduate assistant, Sauer covered a lot of ground, visiting Mennonite as well as Tarahumara settlements, and "discovering" the Parral archives. The archival material from Parral and other local archives yielded two important papers one on aboriginal tribes and languages and the other on aboriginal population (Sauer 1934a, 1935). In effect, the population study was the founding document in the formation of the other "Berkeley school"—that of historical demography. His turn in 1934 to new commitments "back East," which included not only the boardrooms of East Coast foundations but also the corridors of New Deal Washington, especially the new Soil Erosion Service, diverted his Mexican attentions. Still, he managed to spend the summer of 1935 in the field, this time exploring the western edge of the central Mexican plateau, and spending a month in the archives of Mexico City.

Robert Bowman, Isaiah Bowman's son and new Berkeley graduate student joined him in the field. As with Spencer and Hewes, Bowman's summer with Sauer didn't direct him to a follow up season in Mexico, though he did do a study of soil erosion in Puerto Rico for his dissertation (Bowman 1941). Sauer's eastern involvements and administrative duties on campus kept him out of the field for the next three years. By Christmas break 1938-1939 he was back into the field, returning to Sinaloa to visit on-going archaeological excavations by Gordon Ekholm and by Isabel Kelly. That summer he concentrated his attention on Colima, investigating human-plant interactions, the archaeogeography of the region, and historical materials. The results are recorded in his historical monograph, Colima of New Spain in the Sixteenth Century (1948). Henry Bruman, one of his students working on aboriginal beverages, joined him for part of the Colima season.

The accumulated wear and tear from more than a decade of annual trips to Mexico caught up with Sauer after the summer of 1939. Much of the 1939—1940 academic year Sauer was taken up recovering from amoebic dysentery and other undiagnosed maladies. By early 1941 he was well enough to take another sabbatical leave, and once again it was in Mexico. This time the focus was on plant use and craft production in west-central Mexico, as well as a month spent in the Mexico City archives. No direct publications resulted from this season, but Sauer's (1941) seasoned essay "The personality of Mexico" appeared during this time. It has become something of a testament to the benefits and rewards of getting to know a place, region, or nation through periodic immersion. And his growing interest in plant-human relations, observed first-hand in Mexico, provided a base for his seminal Agricultural Origins and Dispersals (1952). The other base for this book was his Rockefeller Foundation sponsored six-month tour of western South America in 1942 (West 1982).

As his research assistant, he took his son Jonathan, at the time a geography graduate student at the University of Wisconsin. Save for an Office of Naval Research-related trip in 1952 to the Caribbean, this South American transect was Sauer's only deflection from his multi-year devotion to fieldwork in many regions of Mexico. The war years 1942-1944 kept Sauer desk bound more than was his custom, but he did return to Mexico at the end of 1944 on a Rockefeller grant, and during the winter and spring of 1945 he was back to familiar locales as well as new ones. He spent a month in Oaxaca and Puebla, parts of Mexico he hadn't previously visited. Following WWII Sauer went back to his annual trips to Mexico. Just as he had pushed farther south into Latin America in increments over two decades, he had recalibrated his estimates of the time depths of various cultural historical processes. He had become convinced that human occupancy of the Americas was far earlier that consensus opinion allowed, and that Mexico's arid landscapes might yield new evidence.

With a 1946 spring leave from teaching, Sauer and the archaeologistEmil Haury, made a transect of Sonora looking for signs of early humans. That summer he went on a similar search down the Baja Peninsula with graduate students William Massey (archaeology) and Edwin Hammond (physical geography). Sites were found that Sauer interpreted as encouraging, and returned in early 1947 with an interdisciplinary group of Berkeley faculty and students under the auspices of the newly organised "Associates of Tropical Biogeography." Sauer's agenda was not necessarily shared, and the faculty geologist, botanist, and paleontologist all preferred to pursue their own interests. That summer he took the first of several trips in which Mexico was traversed from north to south, looking for early human sites, but also of course, other things that caught his eye.

The group included Berkeley colleagues Sherbourne Cook and Leslie Simpson, Jonathan Sauer (now working with Edgar Anderson at Washington University), and two young graduate students, Jesse Walker and Rees Walker (not related). In the summer of 1948 Sauer performed a repeat of the previous year's transect, and again made it as far south as Chiapas, but turned back due to road conditions. Along on the trip were James Parsons, recent Berkeley PhD and graduate student David Lowenthal. Sauer's last trip to Baja California was during another sabbatical leave, in the winter of 1949. This time he eschewed faculty colleagues and took three graduate students, Brigham Arnold, Homer Aschmann, and Thomas Pagenhart. Arnold (1954) settled on a dissertation topic that carried forward Sauer's pursuit of early human occupancy, while

Aschmann (1954) wrote his dissertation on the historical demography and ecology of Central Baja. Pagenhart demurred, filing a dissertation two decades later on historical California water issues. Sauer's final "grand tour" in the summer of 1949 was with graduate students Douglas Powell and Clyde Patton.

Among other objectives were to drive on to Guatemala, but the road ended at the border, and rail transport of the truck was not possible. His final field excursion to Mexico was the summer of 1950. The objective was to study the Pleistocene history of Parras Basin, Coahuila, searching for early human sites, and to find dissertation topics for graduate students Philip Wagner and John Vann. Once they were in place working on the vegetation history and geomorphology and of the basin, Sauer joined the pine expert Nicholas Mirov and Pagenhart on their way south to study pine forests. Among these latter students, only Vann (1959) went on to develop a dissertation out of the trip. Vann was also the last of Sauer's students to file a dissertation on Mexico, save Leonard Sawatzky (1967), who studied Mennonite colonisation, and filed his a decade after Sauer's retirement.

Among Sauer's students who made Mexico a regular destination, Robert West, Donald Brand, and Campbell Pennington stand out. West's dissertation (1948) was on the Parral mining district, but he had "discovered" the archives independently while conducting fieldwork for his master's thesis in geography at UCLA. As he remarked to me (personal communication, 25 May, 1989), Sauer had not so subtly tried to dissuade him from "mining" these archives (presumably so Sauer could do it himself), but West stuck to his guns, and produced a model study. West made over thirty field trips to Mexico from 1936 to 1990, or a span of more than fifty years. His last book on Mexico, Sonora, Its Geographical Personality (1993), distilled a half-century of observations of people and place. Brand (1951; 1966) also made Mexico his primary locus and focus, from 1930 until after retirement in late 1970s, or almost a half-century.

Pennington's dissertation (1959) was directed by James Parsons, but he was very much a Sauer protege. His field excursions in northern Mexico starting in the 1950s and continuing through the 1970s, often on mule back to record the facts of indigenous language and landscapes, were legendary. They yielded a quartet of books (1963; 1969; 1979; 1980) that are monuments to intensive localised fieldwork. A second group that must be considered in the dynamics of repeat visitation research, aside from the repeat researcher and his or her students that also engage the site, are the researchers inspired by the example, but who may or may not be direct legatees. As such, they might be seen as repeat visit proxies. Perhaps the two best examples are Isabel Kelly and William Doolittle. As a Berkeley anthropology graduate student, Kelly took classes with Sauer, but he was not her advisor.

Doolittle is a fourth generation Sauer descendent (Brown and Mathewson 1999). For both, Mexico became their principle focus of fieldwork and research. Isabel Kelly was the second woman to receive a PhD (1932) in anthropology at Berkeley.

Sauer had become her unofficial mentor in graduate school, and afterwards Kelly became a long-standing figure in the shifting retinue of participants in his Mexican fieldwork.

He helped secure funding for her archaeological excavations in western Mexico, and in turn she focused on questions that furthered Sauer's cultural-historical programme, whether cultural boundaries, plant domestication and use, or early human occupations. In this sense, she represents a particular kind of proxy—one that advances a researcher's programme while remaining in situ over an extended period—in this case over several decades devoted to western Mexican archaeology and culture history.

The other kind of proxy is much more common. This is the researcher, whether directly connected to the initiator or not, who continues a line of research begun by someone else in a particular place or region. In effect, he or she performs repeat visitation geography as a proxy for the principle, though the objectives and methods of the research may shift over time. Much of Doolittle's (1988; 1990) research in Mexico has refined and expanded the programme of archaeogeography that Sauer and Brand initiated in northwest Mexico in the 1930s.

His record of repeat visitations is unequalled, and not likely to be surpassed in the future, as it is still ongoing. He has made more than sixty research trips to Mexico, starting in 1977 and spanning more than thirty years. He has made seventeen trips to Sonora, or about one every other year over this thirty-year period.

What Sauer and Brand began in Baja California in the late 1920s, is being continued by Doolittle as well as other geographers. This kind of reiterative practice can produce minor research traditions resistant to change, but it can also allow for multi-decadal engagements by initial researchers or their proxies—that provide windows on site and situational change. Perhaps one of the most encouraging developments in Latin Americanist geography from the perspective of committed "returnees," is the interest being shown by a new generation of Latin American geographers, especially in Mexico and Colombia, in the work of Sauer and his students. Berkeley school classics are being translated, disseminated, and one assumes, read. It is probably too early to tell, but chances are that a new cadre of proxies is being formed in this case, researchers who are not distant "returnees," but ones who culturally at least, have always already been half way or more there.

APPLIED GEOMORPHOLOGY

Geomorphology has traditionally focused on the study of landforms and on the processes involved in their formation. Applied geomorphology is the practical application of this study to a range of environmental issues, both in terms of current problems and of future prediction. Applied geomorphology provides a strategic tool for informed decision-making in public policy development and in environmental resource management. Key areas of application include specific environmental settings, such as the coastal zone or dryland environments; the impacts of land use and management practice on Earth surface processes; and areas susceptible to natural hazards.

Over 60 per cent of the world's population live in the coastal zone in environments ranging from coral atolls, reclaimed or natural wetlands, dune-backed beaches, and barrier islands to cliff tops. Settlements under threat from coastal erosion and flooding from storm events, sea surges, and rising sea level lobby for protective engineering measures to prevent loss of property, livelihood, and life. Geomorphology has several applications in settings of this type.

An understanding of coastal landforms and the processes acting upon them can be used to map areas at risk from cliff failure, beach erosion, and flooding. This approach is of interest to potential developers and the insurance industry and is an important tool in environmental impact assessment. An understanding of the geomorphology of the coastal zone can also be used to predict the effects of modifying the coastal system. The installation of groynes, breakwaters, or protective sea walls has knock-on effects on the natural circulation of water and sediment in the near-shore environment. Artificially stabilizing cliffs to prevent erosion may seem the obvious solution for cliff-top dwellers, but a geomorphological evaluation might predict that this approach could starve beaches of the sediment provided by natural cliff fall, with a consequent impact on longshore drift of sediment, and would relocate the focus of erosion further along the coast.

The nature of the problem may thus change from cliff failure at one site to beach erosion and subsequent flooding at another. An understanding of the nature and complexity of coastal dynamics is thus an essential component of a coastal-zone management strategy and is important in predicting the future effects on coastal landforms of a rise in sea level. River-management strategies for flood alleviation have often adopted engineering solutions concentrated in particular river reaches, which are usually in areas of urban development. Reach-specific intervention measures include lining the natural channel with concrete to prevent erosion and bank instability, channel straightening to force flood water to flow rapidly through particular reaches, and flow-control structures such as sluice gates and reservoirs to control water level. These artificial measures are not always successful in preventing flooding and erosion within the river catchment, and natural sections further downstream may be overwhelmed by the river at peak flood.

The engineered reaches of rivers often become a sterile landscape because fast-flowing water in a concrete-lined channel, with minimal variation in water depth and channel cross-section, provides a poor habitat for wetland flora and fauna. Geomorphology has been applied to 'river restoration' to recreate an integrated river management strategy within artificially created river systems, maximizing biodiversity while controlling river-flow conditions.

Applied geomorphology uses a holistic approach to river response at a catchment-wide scale; the basis here is an understanding of the relationships between river form and process, sediment transport, and the important role of river-bank vegetation. Certain landscapes have specific properties that impinge on our use and development of the environment. In cold environments, the presence of ground ice leads to problems in construction, communication, and housing. In permafrost zones, the ground is permanently frozen except for the upper layers of the soil, which thaw in the summer.

The upper soil, known as the active layer, is subject to repetitive cycles of freezing and thawing, making it geomorphologically active. The ground within the active layer will suffer heaving and deformation, disrupting communications and making road construction impracticable. Applied geomorphology can be used in mapping the active layer and ground ice in areas with differing rocks and sediments.

This information is then used to evaluate the problems that are likely to affect these areas. Ground heaving depends on the depth of the active layer and the type of sediment present; fine-grained silts present more of a problem than gravels. Additional problems in permafrost areas, as, for example, in some regions of Alaska, occur where structures have suffered dramatic subsidence as a result of heating in buildings.

Without appropriate insulation, heat radiates downwards from the building into the ground, thaws the underlying ice, and increases the depth of the active layer, thus effectively changing the structure of the soil. Applied geomorphology is consequently essential in land-use planning and site evaluation, in order to recognize such potential problems as land subsidence, slope instability, invasion of windblown sand, and impacts on natural drainage systems.

Land used for agricultural production may suffer from degradation and desertification as a result of soil erosion, landsliding, and over-extraction of water for irrigation. Much agricultural practice focuses on maximizing yield and profit, often using techniques that can be detrimental to the environment, both in the short and the long term.

Applied geomorphology uses an understanding of the relationships between surface conditions, climate, vegetation, and soil erosion to advise farmers and politicians on how to improve land management for sustainable use of land and water resources. Natural hazards such as volcanic eruptions, earthquakes, and mudflows present a significant risk to the population of the surrounding area. Geomorphological mapping can be used to assess the present condition of the landscape and provide a hazard map.

The expression of a disaster may result in one settlement having significantly different risk assessment. For example, a volcanic eruption may pose a threat from volcanic ash and lava flows, pyroclastic flows, and bombardment from superheated volcanic bombs or associated hazards such as mudflows, depending on topography, soil cover, type of eruption, and predominant wind direction.

This application of geomorphological analysis is of significant interest to the emergency services and the insurance industry. Applied geomorphology can be used in modelling change to landforms and surface processes.

This can include change from human impact on the environment to future prediction of climate change, from short-term El Niño and tropical storm events to longer-term change resulting from greenhouse warming and rising sea levels. In this way, applied geomorphology has a key role in managing the environment to minimize potential degradation of land, water, and natural resources.

GEOMORPHOLOGY-REAL-LIFE APPLICATIONS

Subsidence

Subsidence refers to the process of subsiding (settling or descending), on the part of either an air column or the solid earth, or, in the case of solid earth, to the resulting formation or depression. Subsidence in the atmosphere is discussed briefly in the entry Convection. Subsidence that occurs in the solid earth, known as geologic subsidence, is the settling or sinking by a body of rock or sediment. (The latter can be defined as material deposited at or near Earth's surface from a number of sources, most notably preexisting rock.)

As noted earlier, many geomorphologic processes can be caused either by nature or by human beings. An example of natural subsidence takes place in the aftermath of an earthquake, during which large areas of solid earth may simply drop by several feet. Another example can be observed at the top of a volcano some time after it has erupted, when it has expelled much of its material (i.e., magma) and, as a result, has collapsed.

Natural subsidence also may result from cave formation in places where underground water has worn away limestone. If the water erodes too much limestone, the ceiling of the cave will subside, usually forming a sinkhole at the surface. The sinkhole may fill with water, making a lake; the formation of such sinkholes in many spots throughout an area (whether the sinkholes become lakes or not), is known as karst topography.

In places where the bedrock is limestone—particularly in the sedimentary basins of rivers—karst topography is likely to develop. The United States contains the most extensive karst region in the world, including the Mammoth cave system in Kentucky. Karst topography is very pronounced in the hills of southern China, and karst landscapes have been a prominent feature of Chinese art for centuries. Other extensive karst regions can be found in southern France, Central America, Turkey, Ireland, and England.

Man-made Subsidence

Man-made subsidence often ensues from the removal of groundwater or fossil fuels, such as petroleum or coal. Groundwater removal can be perfectly safe, assuming the area experiences sufficient rainfall to replace, or recharge, the lost water. If recharging does not occur in the necessary proportions, however, the result will be the eventual collapse of the aquifer, a layer of rock that holds groundwater.

In so-called room-and-pillar coal mining, pillars, or vertical columns, of coal are left standing, while the areas around them are extracted. This method maintains the ceiling of the "room" that has been mined of its coal. After the mine is abandoned, however, the pillar eventually may experience so much stress that it breaks, leading to the collapse of the mined room. As when the ceiling of a cave collapses, the subsidence of a coal mine leaves a visible depression above ground.

Uplift

As its name implies, uplift describes a process and results opposite to those of subsidence. In uplift the surface of Earth rises, owing either to a decrease in downward force or to an increase in upward force. One of the most prominent examples of uplift is seen when plates collide, as when India careened into the southern edge of the Eurasian landmass some 55 million years ago. The result has been a string of mountain ranges, including the Himalayas, Karakoram Range, and Hindu Kush, that contain most of the world's tallest peaks.

Plates move at exceedingly slow speeds, but their mass is enormous. This means that their inertia (the tendency of a moving object to keep moving unless acted upon by an outside force) is likewise gargantuan in scale. Therefore, when plates collide, though they are moving at a rate equal to only a few inches a year, they will keep pushing into each other like two automobiles crumpling in a head-on collision. Whereas a car crash is over in a matter of seconds, however, the crumpling of continental masses takes place over hundreds of thousands of years.

When sea floor collides with sea floor, one of the plates likely will be pushed under by the other one, and, likewise, when sea floor collides with continental crust, the latter will push the sea floor under. This results in the formation of volcanic mountains, such as the Andes of South America or the Cascades of the Pacific Northwest, or volcanic islands, such as those of Japan, Indonesia, or Alaska's Aleutian chain.

Isostatic Compensation

In many other instances, collision, compression, and extension cause uplift. On the other hand, as noted, uplift may result from the removal of a weight. This occurs at the end of an ice age, when glaciers as thick as 1.9 mi. (3 km) melt, gradually removing a vast weight pressing down on the surface below.

This movement leads to what is called isostatic compensation, or isostatic rebound, as the crust pushes upward like a seat cushion rising after a person is longer sitting on it. Scandinavia is still experiencing uplift at a rate of about 0.5 in. (1 cm) per year as the after-effect of glacial melting from the last ice age. The latter ended some 10,000 years ago, but in geologic terms this is equivalent to a few minutes' time on the human scale.

Islands

Geomorphology, as noted earlier, is concerned with landforms, such as mountains and volcanoes as well as larger ones, including islands and even continents. Islands present a particularly interesting area of geomorphologic study. In general, islands have certain specific characteristics in terms of their land structure and can be analysed from the standpoint of the geosphere, but particular islands also have unique ecosystems, requiring an interdisciplinary study that draws on botany, zoology, and other subjects. In addition, there is something about an island that has always appealed to the human imagination, as evidenced by the many myths, legends, and stories about islands. Some examples include Homer's *Odyssey,* in which the hero Odysseus visits various islands in his long wanderings; Thomas More's *Utopia,* describing an idealized island republic; *Robinson Crusoe,* by Daniel Defoe, in which the eponymous hero lives for many years on an island with no companion but the trusty native Friday; *Treasure Island,* by Robert Louis Stevenson, in which the island is the focus of a treasure hunt; and Mark Twain's *Adventures of Huckleberry Finn,* depicting Jackson Island in the Mississippi River, to which Huckleberry Finn flees to escape "civilization."

One of the favourite subjects of cartoonists is that of a castaway stranded on a desert island, a mound of sand with no more than a single tree. Movies, too, have long portrayed scenarios, from the idyllic to the brutal, that take place on islands, particularly deserted ones, a notable example being *Cast Away* (2000).

A famous line by the English poet John Donne (1572-1631) warns that "no man is an island," implying that many wish they could enjoy the independence suggested by the concept of an island. Within the Earth system, however, nothing is fully independent, and, as we shall see, this is certainly the case where islands are concerned.

The Islands of Earth

Earth has literally tens of thousands of islands. Just two archipelagos (island chains), those that make up the Philippines and Indonesia, include thousands of islands each.

While there are just a few dozen notable islands on Earth, many more dot the planet's seas and oceans. The largest are these:

- Greenland (Danish, northern Atlantic): 839,999 sq. mi.(2,175,597 sq km)
- New Guinea (divided between Indonesia and Papua New Guinea, western Pacific): 316,615 sq. mi. (820,033 sq km)
- Borneo (divided between Indonesia and Malaysia, western Pacific): 286,914 sq. mi. (743,107 sq km)
- Madagascar (Malagasy Republic, western Indian Ocean): 226,657 sq. mi. (587,042 sq km)
- Baffin (Canadian, northern Atlantic): 183,810 sq. mi. (476,068 sq km)
- Sumatra (Indonesian, northeastern Indian Ocean): 182,859 sq. mi. (473,605 sq km).

The list could go on and on, but it stops at Sumatra because the next-largest island, Honshu (part of Japan), is less than half as large, at 88,925 sq. mi. (230,316 sq km). Clearly, not all islands are created equal, and though some are heavily populated or enjoy the status of independent nations (e.g., Great Britain at number eight or Cuba at number 15), they are not necessarily the largest. On the other hand, some of the largest are among the most sparsely populated.

Of the 32 largest islands in the world, more than a third are in the icy northern Atlantic and Arctic, with populations that are small or practically nonexistent.

Greenland's population, for instance, was just over 59,000 in 1998, while that of Baffin Island was about 13,200. On both islands, then, each person has about 14 frozen sq. mi. (22 sq km) to himself or herself, making.

Continents, Oceans, and Islands

Australia, of course, is not an island but a continent, a difference that is not related directly to size. If Australia *were* an island, it would be by far the largest. Australia is regarded as a continent, however, because it is one of the principal landmasses of the Indo-Australian plate, which is among a handful of major continental plates on Earth.

Whereas continents are more or less permanent (though they have experienced considerable rearrangement over the eons), islands come and go, seldom lasting more than 10 million years. Erosion or rising sea levels remove islands, while volcanic explosions can create new ones, as when an eruption off the coast of Iceland resulted in the formation of an island, Surtsey, in 1963.

Islands are of two types, continental and oceanic. Continental islands are part of continental shelves (the submerged, sloping ledges of continents) and may be formed in one of two ways.

Rising ocean waters either cover a coastal region, leaving only the tallest mountains exposed as islands or cut off part of a peninsula, which then becomes an island. Most of Earth's significant islands are continental and are easily spotted as such, because they lie at close proximity to continental landmasses. Many other continental islands are very small, however; examples include the barrier islands that line the East Coast of the United States. Formed from mainland sand brought to the coast by rivers, these are technically not continental islands, but they more clearly fit into that category than into the grouping of oceanic islands.

Oceanic islands, of which the Hawaiian-Emperor island chain and the Aleutians off the Alaskan coast are examples, form as a result of volcanic activity on the ocean floor. In most cases, there is a region of high volcanic activity, called a hot spot, beneath the plates, which move across the hot spot. This is the situation in Hawaii, and it explains why the volcanoes on the southern islands are still active while those to the north are not: the islands themselves are moving north across the hot spot. If two plates converge and one subducts, a deep trench with a parallel chain of volcanic islands may develop. Exemplified by the Aleutians, these chains are called island arcs.

Island Ecosystems

The ecosystem, or community of all living organisms, on islands can be unique owing to their separation from continents. The number of life-forms on an island is relatively small and can encompass some unusual circumstances compared with the larger ecosystems of continents. Ireland, for instance, has no native snakes, a fact "explained" by the legend that Saint Patrick drove them away. Hawaii and Iceland are also blessedly free of serpents.

Oceanic islands, of course, tend to have more unique ecosystems than do continental islands. The number of land-based animal life-forms is necessarily small, whereas the varieties of birds, flying insects, and surrounding marine life will be greater owing to those creatures' mobility across water. Vegetation is relatively varied, given the fact that winds, water currents, and birds may carry seeds.

Nonetheless, ecosystems of islands tend to be fairly delicate and can be upset by the human introduction of new predators (e.g., dogs) or new creatures to consume plant life (e.g., sheep). These changes sometimes can have disastrous effects on the overall balance of life on islands. Overgrazing may even open up the possibility of erosion, which has the potential of bringing an end to an island's life.

2

Concept of Geomorphic Processes

CONCEPT

The surface of Earth is covered with various landforms, a number of which are discussed in various entries throughout this book. This essay is devoted to the study of landforms themselves, a subdiscipline of the geologic sciences known as geomorphology. The latter, as it has evolved since the end of the nineteenth century, has become an interdisciplinary study that draws on areas as diverse as plate tectonics, ecology, and meteorology. Geomorphology is concerned with the shaping of landforms, through such processes as subsidence and uplift, and with the classification and study of such landforms as mountains, volcanoes, and islands.

An Evolving Area of Study: Geomorphology is an area of geology concerned with the study of landforms, with the forces and processes that have shaped them, and with the description and classification of various physical features on Earth. The term, which comes from the Greek words *geo,* or "Earth," and *morph,* meaning "form," was coined in 1893 by the American geologist William Morris Davis (1850-1934), who is considered the father of geomorphology.

During Davis's time, geomorphology was concerned primarily with classifying different structures on Earth's surface, examples of which include mountains and islands, discussed later in this essay. This view of geomorphology as an essentially descriptive, past-oriented area of study closely aligned with historical geology prevailed throughout the late nineteenth and early twentieth centuries.

By the mid-twentieth century, however, the concept of geomorphology inherited from Davis had fallen into disfavor, to be replaced by a paradigm, or model, oriented toward physical rather than historical geology. (These two principal branches of geology are concerned, in the first instance, with Earth's past and the processes that shaped it and, in the second instance, with Earth's current physical features and the processes that continue to shape it.)

Rethinking Geomorphology: As reconceived in the 1950s and thereafter, geomorphology became an increasingly exact science. As has been typical of many sciences in their infancy, early geomorphology focused on description rather than prediction and tended to approach its subject matter in a qualitative fashion. The term qualitative suggests a comparison between qualities that are not defined precisely, such as "fast" and "slow" or "warm" and "cold." On the other hand, a quantitative approach, as has been implemented for geomorphology from the mid-twentieth century onward, centers on a comparison between precise quantities—for instance, 10 lb. (4.5 kg) versus 100 lb. (45 kg) or 50 MPH (80.5 km/h) versus 120 MPH (193 km/h).

As part of its shift in focus, geomorphology began to treat Earth's physical features as systems made up of complex and ongoing interactions. This view fell into line with a general emphasis on the systems concept in the study of Earth. As geomorphology evolved, it became more interdisciplinary, as we shall see. This, too, was part of an overall trend in the earth sciences toward an approach that viewed subjects in broad, cross-disciplinary terms as opposed to a narrow focus on specific areas of study.

Landforms and Processes: Two concerns are foremost within the realm of geomorphology, and these concerns reflect the stages of its history. First, in line with Davis's original conception of geomorphology as an area of science devoted to classifying and describing natural features, there is its concern with topography. The latter may be defined as the configuration of Earth's surface, including its relief (elevation and other in equalities) as well as the position of physical features.

These physical features are called landforms, examples of which include mountains, plateaus, and valleys. Geomorphology always has involved classification, and early scientists working in this subdiscipline addressed the classification of landforms. Other systems of classification, however, are not so concerned with cataloging topographical features themselves as with differentiating the processes that shaped them. This brings us to the other area of interest in geomorphology: the study of how landforms came into being.

Shaping the Earth: Among the processes that drive the shaping of landforms is plate tectonics, or the shifting of large, movable segments of lithosphere (the crust and upper layer of Earth's mantle). Plate tectonics is discussed in detail within its own essay and more briefly in other areas throughout this book, as befits its status as one of the key areas of study in the earth sciences.

Other processes also shape landforms. Included among these processes are weathering, the breakdown of rocks and minerals at or near the surface of Earth due to physical or chemical processes; erosion, the movement of soil and rock due to forces produced by water, wind, glaciers, gravity, and other influences; and mass wasting or mass movement, the transfer of earth material, by processes that include flow, slide, fall, and creep, down slopes. Also of interest are fluvial and eolian processes (those that result from water flow and wind, respectively) as well as others related to glaciers and coastal formations.

Human activity also can play a significant role in shaping Earth. This effect may be direct, as when the construction of cities, the building of dams, or the excavation of mines alters the landscape. On the other hand, it can be indirect. In the latter instance, human activity in the biosphere exerts an impact, as when the clearing of forest land or the misuse of crop land results in the formation of a dust bowl.

Interdisciplinary Studies: As noted earlier, geomorphology is characteristic of the earth sciences as a whole in its emphasis on an interdisciplinary approach. As is true of earth scientists in general, those studying landforms and the processes that shape them do not work simply in one specialty. Among the areas of interest in geomorphology are, for example, deep-sea geomorphology, which draws on oceanography, and planetary geomorphology, the study of landscapes on other planets. When studying coastal geomorphology, a geologist may draw on realms as diverse as fluid mechanics (an area of physics that studies the behavior of gases and liquids at rest and in motion) and sedimentology. The investigation of such processes as erosion and mass wasting calls on knowledge in the atmospheric sciences as well as the physics and chemistry of soil. It is almost inevitable that a geomorphologic researcher will draw on geophysics as well as on such subspecialties as volcanology. These studies may go beyond the "hard sciences," bringing in such social sciences as geography.

GEOLOGICAL TEXTURE

Texture (Geology)

Texture in geology refers to the physical appearance or character of a rock, such as grain size, shape, arrangement, and pattern at both the megascopic or microscopic surface feature level. This includes the geometric aspects and relations amongst the component particles or crystals which is called the crystallographic texture or preferred orientation. The term microstructure is preferred by some petrographers. Textures can be quantified in many ways. The most common parameter is the crystal size distribution.

Igneous Textures

Igneous textures include the rock textures occurring in igneous rocks. Igneous textures are used by geologists in determining the mode of origin igneous rocks and are used in rock classification. There are six main types of textures; phaneritic, aphanitic, porphyritic, glassy, pyroclastic and pegmatitic.

Aphanitic (*a* = not, *phaner* = visible) rocks in contrast to phaneritic rocks, typically form from lava which crystallize rapidly on or near the Earth' surface. Because extrusive rocks make contact with the atmosphere they cool quickly, so the minerals do not have time to form large crystals. The individual crystals in an aphanitic igneous rock are not distinguisable to the naked eye. Examples of aphanitic igneous rock include basalt, andesite and rhyolite.

Glassy or *vitreous* textures occur during some volcanic eruptions when the lava is quenched so rapidly that crystallization cannot occur. The result is a natural amorphous glass with few or no crystals. Examples include obsidian and pumice.

Pegmatitic texture occurs during magma cooling when some minerals may grow so large that they become massive (the size ranges from a few centimetres to several metres). This is typical of pegmatites.

Phaneritic (*phaner* = visible) textures are typical of intrusive igneous rocks, these rocks crystallized slowly below the Earth's surface. As a magma cools slowly the minerals have time to grow and form large crystals. The minerals in a phaneritic igneous rock are sufficiently large to see each individual crystal with the naked eye. Examples of phaneritic igneous rocks are gabbro, diorite and granite.

Porphyritic textures develop when conditions during cooling of a magma change relatively quickly. The earlier formed minerals will have formed slowly and remain as large crystals, whereas, sudden cooling causes the rapid crystallization of the remainder of the melt into a fine grained (aphanitic) matrix. The result is an aphanitic rock with some larger crystals (phenocrysts) imbedded within its matrix. Porphyritic texture also occurs when magma crystallizes below a volcano but is erupted before completing crystallization thus forcing the remaining lava to crystallize more rapidly with much smaller crystals.

Pyroclastic (*pyro* = igneous, *clastic* = fragment) textures occur when explosive eruptions blast the lava into the air resulting in fragmental, typically glassy material which fall as volcanic ash, lapilli and volcanic bombs.

Geologic Map Patterns for Canvas

The default installation of drawing programs such as Canvas (R) or Illustrator (R) provide insufficient patterns for designating areas on geologic maps. Although these programs make it quite simple to define a custom pattern for further use, generally this functionality is not used because of time constraints and lack of artistic talent. The USGS recognizes this and-in an attempt to standardize-provides a multitude of geologic map symbols and patterns in its Open File Report 99-430 (link to USGS publication at USGS site). The map patterns are professionally designed in Adobe Illustrator(R) format und provided by the USGS in EPS (Illustrator version) and PDF (overview) formats. Here we offer these patterns for download in Adobe Illustrator (R), Deneba Canvas (R), OpenOffice, EPS (generic version) and PDF (single fill element version) formats.

After downloading and opening the appropriate file, the map patterns become automatically available in the pattern swatch in Illustrator (R). In Canvas (R), open the file and tear away the object fill item in the tool box. Then choose the Symbol tab and expand the ink manager with the downward pointing triangle. Just drag a pattern into the square symbol definition area to make it available as a fill pattern for general use. Here is a hint from John Hunt for Freehand (R) users:

First download the appropriate PDFs from the web. Then Import the file patterns you want into Freehand (do NOT use OPEN). It then appears (be patient!) into Freehand as many individual patterns. Select the pattern tile you want. Then go to the menu Window/Inspectors/Fill which will bring a floating pallette onto your worktop. Select Tiles from the pull down menu, and hit Paste In.

Adobe Illustrator (R) version 7 or greater and Canvas (R) version 7 or greater are required. OpenOffice 1.0 can load the provided files but it does not have the notion of pattern fills. It is possible to emulate the effect by merging and intersecting shapes.

PRINCIPLES OF TERRAIN ANALYSIS

Digital terrain models have found wide applications in disciplines such as mapping, remote sensing, civil engineering, mining engineering, geology, geomorphology, military engineering, land planning, and communications. Digital Terrain Modeling: Principles and Methodology is the only authored book to provide comprehensive coverage of recent developments in the field. The volume includes an examination of data acquisition and a review of the presentation of DTMs in databases, in contour form, and in various forms of computer graphics. It explores the theories, methods, and algorithms of digital terrain modeling, and focuses on quality control and accuracy. This book discusses extensions of digital terrain models for specific problems and anticipates where this research will lead. Topics include terrain analysis, sampling strategy, acquisition methodology, surface modeling principles, triangulation algorithms, interpolation techniques, on-line and off-line quality control in data acquisition, DTM accuracy assessment and mathematical models for DTM accuracy prediction, multi-scale representation, data management, contouring, visual analysis (or visualization), the derivation of various types of terrain parameters, and future development and applications.

Terrain Analysis Considerations

Proper terrain analysis is as essential to success on the battlefield today as it was when Frederick the Great ruled Prussia over 200 years ago. It is especially relevant when a force must attack an enemy which has had time to prepare a well-organized defense. This is exactly the situation most units training at U.S. Army Combat Training Centers (CTCs) must confront. Recent CTC trends indicate that units are having difficulty conducting successful combined arms breaching operations. Staff planners can help reverse that trend by mastering categories of terrain analysis tasks which foster mobility and breach operation planning. Today's planners use modern automated tools and equipment to achieve very detailed analyses of the terrain and weather. The utility and availability of these tools continues to increase, and they provide timely support to time-strapped planners.

The Planner's Role

Most staff planners at the brigade level and below, especially the intelligence officer, assistant brigade engineer and task force engineer, realize the important role they have in conducting terrain analysis tasks in support of maneuver and breach operations planning. All BOS (Battlefield Operating Systems) representatives must consider the terrain's effects, in that the maneuver S-2 must work with Engineer, Armor, Field Artillery, Infantry, Air Defense and others to properly template and assess friendly avenues of approach, enemy obstacles and enemy battle positions (BPs). These BOS planners must survey the terrain for areas which may be used for support-by-fire (SBF) positions, attack-by-fire (ABF) positions, any covered and concealed routes into these positions, and other locations important in the planning process.

Once planners complete the initial mission analysis/IPB (Intelligence Preparation of the Battlefield), many planners think that their role as terrain analysts/integrators is complete. The reality is that the planner must perform *continuous* analysis through the entire planning, preparation and execution phases. Planners and units must be prepared to modify their plan based on this continuous evaluation of data and the situation. Planners must continually and rapidly analyze, integrate and disseminate validated enemy obstacle and dispositional intelligence. CTC trends also show a reluctance of units to modify their plans, even when indicators are present that should prompt them to make adjustments. The engineer bears considerable responsibility for this process as the force's primary expert on the terrain and its effects.

Commanders and staffs should be prepared to incorporate adjustments to the scheme of maneuver plan even during the execution of the plan, provided the recommended change makes timely, tactical sense. For instance, if R & S (Reconnaissance and Surveillance) reveals a previously unknown weakness in the enemy frontage or obstacle system, or notes a previously unidentified obstacle system on a planned route, the staff should have already built in the flexibility needed to quickly adjust the original plan.

Categories of Terrain Analysis

A TTP (tactic, technique and procedure) for grouping the terrain analysis tasks associated with planning for breaching operations places the tasks into three general categories: tasks that support template analysis, tasks that support maneuver analysis, and tasks that support planning for the breach force and assault force.

The first category, "Template Analysis" tasks, includes tasks associated with templating enemy obstacles and BPs.

The second category, "Maneuver Analysis" tasks, deals with those tasks that assist maneuver units in selecting avenues of approach (AAs) and positions which best support accomplishing their assigned tasks. It also uses information on enemy obstacles, positions, and the terrain to select the best area(s) for the point of breach (POB) and point of penetration (POP).

The final category, "Breach Analysis" tasks, involves unit-level planning for placement of vehicles and the assets used to conduct the POB and POP missions. We will examine each of these categories in more detail to emphasize the important role that planners have in influencing the decision to move and breach in the most advantageous area(s).

Terrain Task Categories

Template Analysis: The template analysis provides a doctrinal basis for estimating enemy disposition and alignment. The purpose of the tasks associated with Template Analysis is to develop an accurate situational template (SITTEMP). In creating the SITTEMP, the doctrinal template is fitted to the situation. Validated intelligence of the actual enemy composition, disposition and intent, confirmed obstacle effort, and the effects of terrain and weather help predict the most likely COA. Each separate staff planner brings a particular specialty to the table to assess and estimate their enemy counterpart's capabilities, limitations and intentions, and assists in developing several potential enemy COAs.

Planners begin by analyzing the natural and manmade features of the terrain (roads, cities, vegetation, streams, hills) and the first three military aspects of terrain (the O-C-O in OCOKA) to create a COO (Combined Obstacles Overlay). This product reflects the physical effects of the terrain and weather on the battlefield, particularly with respect to mobility. This overlay is the base product upon which the planner builds the MCOO (Modified Combined Obstacles Overlay). The planner then adds the assessments from the MCOO to the doctrinal template to create the situational template.

The engineer planner plays a key role in this process. His knowledge of obstacle tactics, the effects of terrain and weather, enemy obstacle composition, and enemy engineer capabilities assists the S-2 in accurately *estimating* enemy countermobility and survivability effort and *templating* its most likely location. Each BOS element adds similar input to the SITTEMP.

Terrain teams are especially valuable in this process, as each unit's supporting terrain team creates a detailed COO for each AO and AI. Many planners are unfamiliar with the mission, capabilities, and support provided by terrain teams, as well as how to receive or request this support. Terrain team products enable friendly forces to maximize the use of the terrain in any military operation.

Doctrinal knowledge and existing obstacle intelligence, combined with terrain products created either by the unit or a terrain team, greatly facilitate the planner's ability to select the most likely locations of enemy BPs, as well as the likely visibility and available fields of fire from those BPs. Examples of these products include weapons fan and line-of-sight (LOS) analyses.

A weapons fan product is normally created using digital map or digital image data of the area of operations plus digital elevation data. The product is a graphic depiction of the terrain that is directly visible or not visible from a position on the ground, and can, therefore, be hit with direct fire or is deadspace. The user specifies the location of the weapon, its height above the ground, the target's height above the ground, and how far out from the weapon the fan extends (the maximum effective range). The user can also draw a circle around the weapon at its maximum effective range, thus identifying where the weapon cannot reach its maximum range.

This product shows three weapons fans: two in valleys and one in the center of a broad plain. The fans are drawn with range circles at a maximum effective range of 2,500 meters. The weapons in the valleys can only fire for a short distance before their visibility ends; the center weapon can reach out to its maximum effective range in nearly every direction. Depending on the software used, the product may or may not consider the effects of vegetation on visibility. These products can be useful to any BOS. The product helps to precisely place BPs and obstacle systems for maximum advantage. For instance, the planner can position vehicles to place interlocking direct fire on an obstacle system while using terrain features to shield themselves from direct fire. A unit planner can create this product in minutes.

These products are prepared using common software (TerraBase II), and are used to determine intervisibility between two precise points on the ground.

Another example of template analysis involves knowledge of soil conditions and doctrine. The engineer planner can use detailed knowledge of the soils in the AO combined with doctrinal knowledge of enemy survivability equipment to identify areas where the enemy is likely to construct vehicle fighting positions and the total number that the enemy can realistically dig in the time available. In addition, the engineer, knowing enemy obstacle emplacement equipment and techniques, can also estimate the total amount of enemy obstacle effort available. This information, used with the previously mentioned products and the soils analysis, helps to template where and how the enemy will defend. The SITTEMP is an estimate of how the enemy will defend, and assists in driving R&S planning. As R&S confirms or denies the estimate, the S-2 and engineer refine and adjust the SITTEMP. This requires the staff to continually refine, adjust and synchronize the scheme of maneuver based on the likely enemy defensive posture.

Maneuver Analysis: The next category of terrain analysis tasks provides planning support to COA development. Maneuver analysis helps resolve questions concerning on-road and off-road trafficability, denied routes, the locations of enemy positions or obstacles, and numerous other similar considerations. Maneuver analysis helps determine the optimum locations for support-by-fire (SBF) positions, attack-by-fire (ABF) positions, the POB and POP, hide positions, and indirect fire targeting. The result of maneuver analysis is our most advantageous use of the terrain in the attack.

The terrain and how the weather affects it is a common thread binding each BOS together in the overall staff planning process. Every ground vehicle must move on the same terrain in the AO or AI. Their movement and actions are often constrained to specific AAs or mobility corridors leading to the point of breach. Planners must consider these constraints, for if the terrain or weather allows the combat power to outstrip its resupply, or if combat vehicles are forced into moving too slowly or for too long in open areas, the result can range from a slowed or incomplete mission to unacceptable losses or mission failure.

Since this set of tasks requires analysis of how vehicles will interact with the terrain, it requires a somewhat different focus. The planner can use the MCOO to classify the AAs and cross-mobility corridors as unrestricted,

restricted or severely restricted. The planner can then determine the tactical viability of each AA or corridor, verifying data such as:

- Estimated vehicle/convoy speed.
- Total vehicle capacity (either in a column for a convoy or across a frontage in a maneuver formation).
- Provision for two-way traffic and resupply/rest/repair areas.
- The numbers of natural or manmade obstacles along the route.
- The numbers of gaps or water-crossing sites.
- The route's composition (gravel, dirt, pavement) and its potential for degradation in inclement weather.
- The amount of cover and concealment, where it is good and where it is not.
- The route's proximity or directness to the proposed POB and POP.
- The availability of space for the force to transition from convoy "formation" to attack formation with minimal difficulty.

The SITTEMP and MCOO (created by the engineer/S2), combined with information from unit-created products and specialized products from the terrain team, help to conduct this analysis.

The terrain team produces specific products to support this analysis, including visible area plot, elevation, slope, soil, and route maneuverability/ trafficability products. The cross-country (off-road) and on-road trafficability products are especially useful in this analysis. The planner can create or use weapons fan and line-of-sight plots to identify areas the enemy could use for ambush sites and the likely locations of the ambush vehicles. The planner can use the same analysis process to identify the optimum covered/concealed routes away from the breach site toward the POP (on the far side of the objective).

Planners must conduct an integrated analysis of the weather and terrain using full situational awareness. Rainfall on the generally rocky, nonvegetated terrain at the NTC will have a different set of effects than on the generally wetter, heavily vegetated terrain in Korea or Europe. In addition, planners must separately evaluate each dry or wet gap as a potential obstacle given the composition of the force and the outlook of the weather. When there is no rainfall, maneuver units may be capable of crossing dry or wet gaps without engineer support. As rainfall increases, however, dry gaps become wet and wet gaps can become passable only with bridging assets or ford preparation, requiring bypass if such assets are unavailable. Each gap the force must cross is a drain on time and assets, and requires a significant amount of coordination. When the force must use crossing assets, staff planners must ensure there are sufficient assets to make an uninterrupted series of crossings along the planned route(s).

Breach Analysis

Breach analysis tasks are an extension of maneuver analysis tasks, and involve detailed breach planning at the unit level. The planner must evaluate the routes from the LD through both tactical and protective obstacles. The planner must consider the terrain, the weather, the SITTEMP, current obstacle intelligence, friendly movement capabilities, the optimal sites for ABF and SBF positions, friendly and enemy weapon systems capabilities, and any other information relevant to the operation. All planners must ensure they fully analyze, resource, plan and coordinate each of the breaching fundamentals (Suppress, Obscure, Secure, Reduce, Assault, or SOSR-A) (FM 3-34.2). A breach is a mission shared by many BOSs, and is critically dependent on proper evaluation and use of the terrain and weather and full participation by all BOSs.

At the breach site itself, the planner's role is critical in influencing the selection of the POB. Planners should use the MCOO, the SITTEMP, terrain team products, unit terrain products, the most recent R&S ground intelligence or the most recent imagery (from a plane, UAV or satellite) of the routes and obstacle systems that exists. These products and data allow the planner to get near-real-time data on obstacles, enemy positions, and potential breach sites.

Once the locations of the enemy obstacles and BPs are known, the planner can use the products to identify the terrain most vulnerable to direct fire from enemy BPs. The planner can use IV line, weapons fan plots and select line-of-sight analyses to choose the optimal locations for SBF and ABF positions, the most covered/concealed routes into them, and the POB. The planner can pinpoint enemy direct fire systems and obstacles, check enemy obstintel, and the POB. He can analyze the POB to determine the best assets and how to employ them in the breach drill/execution (breach force unit-level planning). The planner can use reverse LOS analysis or weapons fan plots from each proposed SBF position to identify how many BPs can observe activities at the SBF position and along the obstacle. If the planner knows the locations of individual fighting positions, he can identify exactly which positions are within range of, and are able to observe, the obstacles and SBF positions. In this way, the planner pinpoints locations along the obstacles which may provide cover or concealment from the most direct fire weapons systems. He can select the positions that the breach force's security element can use to provide local security at the breach sites.

Planners also support the breach force commander with vegetation and soil condition overlays that indicate areas that are conducive to the breaching techniques available (mechanical, explosive, manual or electronic). Terrain products indicate the type of terrain in the area (rolling, rocky, soft) and its micro-effects on visibility, movement and breaching. The engineer and S-2 also use the MCOO to identify mobility corridors the enemy could use for a counterattack. Based on these analyses, the planner can recommend the best area(s) for the breach.

Tools

There are two primary tools used in the field to conduct terrain/battlefield evaluation and visualization. The Digital Topographic Support System (DTSS) is the terrain team's primary automated tool used to create terrain products.

TerraBase II is the primary automated PC-based tool used by units at the brigade level and below. Each tool has advantages and limitations, and both are used to give planners optimal battlefield evaluation and visualization.

The DTSS is a set of automation equipment consisting of a powerful CPU, a 24" or 36" roll/drum scanner and several 36" color plotters, although the actual amounts and numbers of equipment on hand will vary between units. The system is generally mounted on a HMMWV, 5-ton, FMTV or other mobile platform. Until recently, terrain teams were only present at the division and corps levels and provided support to their command group and MSCs, so the DTSS was only available at the division level or higher. However, under the Force XXI restructuring initiatives, there is now a DTSS suite and a four-person terrain detachment with each maneuver brigade in the digital corps.

The DTSS is invaluable when units are preparing to enter an area of operations. It is a powerful tool that can create extremely precise products that evaluate all relevant considerations: terrain, weather, soils, hydrology, vegetation, and other factors.

It can create products such as soil analysis and vegetation overlays and a MCOO. These products are usually made well before a unit enters a particular AO. The advantage of the DTSS is that all the input data and output products are digital and can be stored, sent, printed, or later manipulated to fit any particular need. The system also has a limited rapid reproduction capability so that products can be made and printed while in theater. More leaders need to realize the capabilities of this system and use its products in support of the continuous Military Decision-Making Process/Intelligence Preparation of the Battlefield (MDMP/IPB) processes.

Terrain teams normally produce a general suite of products for each mission to support the MDMP. These include a MCOO, separate specialized cross-country mobility, hydrology, lines of communication, elevation/slope tint and soils products. These products are often distributed down to the brigade S-2 in hardcopy format, although local distribution SOP may vary. These products are intended for use by all BOSs; all planners need to know these products exist and ensure they have access to them as soon as they are available. In addition to this general set of products, teams and detachments can provide more detailed and customized products, depending on the situation and request.

There are several different programs directly available to the unit-level planner to conduct terrain analysis tasks, but the most commonly used and available is TerraBase II. TerraBase II displays terrain in 3D perspective views and conducts relief-based analysis. Many types of imagery and digital maps can be draped over a 3D relief of the terrain to provide a very useful representation of the terrain features.

TerraBase II is a freeware program created to support units at the brigade level and below. It is an easy-to-use terrain evaluation tool usable on low-end PCs. The program provides planners from any BOS the ability to create the specific products important to the ground soldier, such as lines of sight, weapons fans and image maps. The Engineer School currently uses TerraBase in engineer officer training, to support the expectation of maneuver commanders that the engineer is the terrain expert.

The program uses National Imagery and Mapping Agency (NIMA) data, United States Geological Survey (USGS) data and many other types of data, including satellite imagery.

The main functions planners use are the weapons fan/range circle and line-of-sight tools. Some additional analysis functions include slope and elevation analyses, terrain categorization, and oblique and perspective views. The program also creates "movies" showing a rotating panoramic view from a specified location, or "fly-throughs" over the terrain using user-identified viewpoint information. The program is limited in that it does not consider the effects of vegetation, hydrology or weather, so the planner must always evaluate the validity and utility of the TerraBase II products based on these unassessed criteria.

Planners can download versions 3.0, 4.0, and 5.0 of the program, training documents including videos on how to execute functions, and a Microsoft Word tutorial from the Internet. Version 3.0 provided the basis for a multimedia CD-ROM TerraBase II tutorial package recently distributed by the U.S. Army Engineer School to each active duty, USAR and ARNG engineer battalion, brigade and group, to each of the tactical branch schools and to the CTCs.

The intent is to provide the field a means to self-train on the program using the package plus any locally reproduced copies of the package. The downloaded programs come with very little digital data, but the tutorial package includes digital elevation, map and imagery data for common military training areas, such as NTC, JRTC and CMTC, plus data for other areas of interest. The CD-ROM is currently given to each student at the engineer basic, advanced and pre-command courses, and once annually to engineer students at CGSC.

The program can display high-resolution imagery, including 1-meter resolution products from NIMA, the USGS or commercial sources. The program displays GeoTIFF images created using high-resolution (1m) color MrSID (Multiresolution Seamless Image Database) imagery. Planners can request this imagery for Fort Irwin, CA, Fort Polk, LA, and Fort Leonard Wood, MO, and use it to create image files usable in TerraBase II. This imagery allows the user to combine the tactical analysis tools of TerraBase with satellite imagery that commanders and staffs often find easier to interpret and more useful than digital map data. Other program uses include, but are not limited to:

- Planning direct fire on/from proposed battle positions.
- Re-transmission position planning.
- Confirming visibility to objectives/NAIs/TAIs.
- Estimating slope along routes for resupply vehicles and others.
- Delineation of air avenues of approach.
- Identifying optimal locations for LZs/DZs/base camps.
- Placing OPs for optimal viewing and concealed movement.
- Placing artillery pieces on low-slope areas.
- Identifying potential ambush sites.
- Conducting 3D movement/mission rehearsals.

TerraBase is free and easy to learn, and anyone in the force can use and apply TerraBase and its products to their missions. However, it is important to know that much of the digital image, map and elevation data used is "Limited Distribution" and only DOD military and approved civilian personnel are authorized to use the data.

The planner must make or request TerraBase II/terrain team products before, during and after the mission. Products received before the mission assist in planning. Products made or received during the mission assist in course correction and mission flexibility. Products made after the mission assist in BDA, analysis of the effectiveness of the maneuver and the breach, and augment after-actions/lessons learned records.

Products

There are three basic types of topographic or terrain visualization products with which the planner needs to be familiar: standard, nonstandard and TerraBase II products.

Standard products are hardcopy or softcopy items which are mass-produced for the DOD and are available through the supply system using an NSN. These include hardcopy or digital data such as maps, digital imagery, elevation data, and other feature data. The hardcopy products are normally paper, while digital products are normally on CD-ROMs.

Nonstandard products also include hardcopy and softcopy items, but they are not mass-produced. Terrain teams create these products using the DTSS and they are tailored to support specific operations. These products graphically illustrate the latest information about an area and serve as tactical decision aids (TDAs). A TDA is a product used to assist planners and commanders in making decisions about operations based on the effects of terrain and weather. An example is a cross-country mobility model illustrating the speed at which a mechanized brigade can move off-road. A day-old satellite image with military grid lines and a legend is much more useful than a 35-year old map.

Some other TDAs planners could use include Terrain Category (TerCat) and elevation tint products, which BOSs, such as Field Artillery and Air Defense Artillery, often use. Planners can use TerCat products to identify low-slope areas from which artillery pieces can fire, while elevation tints are used to identify air AAs. A sample package of nonstandard TDA products created for an operation could include the items listed below; the full range of products is listed in FM 3-34.230:

- Observation and fields of fire.
- Concealment.
- MCOO.
- Key terrain.
- Avenues of approach.
- LOC-roads, bridges, tunnels, sharp curves, steep grades.
- Hydrology-depth, width, velocity, bottom material, bank height, ford sites.
- Facilities-natural resources, manufacturing facilities.
- Image map (may be combined with facilities and key terrain).
- Elevation tint, placed over maps or imagery.
- LZ/DZ.

The third type of product is the TerraBase product created at the unit level. This is simply any product created using any function of TerraBase which is then used for tactical terrain visualization, evaluation and decisionmaking. A U.S. Army Engineer School TTP tool for identifying and collating all the various products necessary or desired for the terrain planner is called the Terrain Mission Folder (TMF) (mentioned in both FM 3-34.2 and FM 3-34.230). This product is a folder or collection of terrain/weather data necessary to support the planner in the decisionmaking process. The TMF is a set of mission-specific topographic products that is focused on a 10km x 10km area of operations, and is created in advance of the operation to support the IPB process. It may include:

- Standard products, either hardcopy or digital.

- Nonstandard products created by a terrain team.
- TerraBase II products specifically created for the operation.
- Intelligence and weather reports.
- Graphics from the Engineer planner.
- Photographs.
- Any other data deemed important to the planner.

Training

Knowledge of the process of terrain, weather and situational analysis and how to leverage the various terrain tools, organizations and products is key to the planner's ability to make the correct recommendation to the commander. Therefore, staff planners must ensure they and their units receive training across the full spectrum of the terrain and weather integration process. The training should occur at home station, and should focus on several areas of interest. This home-station training should include:

- Periodic refreshers on the MDMP and IPB processes, with special focus on terrain analysis. Training should always include a full review and integration of terrain tools, products and area evaluations.
- Unit-or individual-level training on the TerraBase II program, including practical exercises on selected areas of interest. This allows the exercise products to remain at the unit for potential later use during deployments/field training. Where and when possible, each product should be field-tested for validity and this knowledge incorporated into future products.
- The planner should request that the terrain team warrant officer (OIC) brief the unit and demonstrate/show the products the team can provide the planners, when and how the planner should request them and when each product is likely to be available during the planning process. It is important to use a sample of the actual products during this training.
- Staff-exercise coordination should be a part of all training and exercises to ensure information and product flow does not stop at any level or echelon but makes it down to the bottom, user level. The staff-exercise should also ensure each BOS planner incorporates his specialty into the process.

Terrain Analysis

Whether concrete, sand, snow, or jungle, knowing your terrain and the obstacles you may face on a mission can mean all the difference in completing your goals. Terrain analysis software from Intergraph delivers extensive topographical and strategic information. From terrain detection to terrain visualization, Intergraph terrain analysis solutions provide preparatory support for mission planning systems and beyond. Whether preparing for an environmental engineering project, creating emergency response simulation, or devising new supply delivery route options, terrain analysis tools from Intergraph provide the flexibility and power that your military or intelligence agency needs.

- Terrain Analysis Software: From terrain detection to terrain visualization and more, learn more today about how our software solutions work together to deliver comprehensive terrain analysis tools.
- Contact Intergraph: If you are interested in learning more about the specific ways Intergraph terrain analysis software can benefit your enterprise, please contact us via our information request form today. Additionally, to stay current on technological advancements in terrain analysis, terrain detection, terrain visualization, and more, simply sign up for our e-Connection newsletter.

Terrain Analysis Solutions

As terrain differs depending on geography and objectives, Intergraph offers a wide selection of terrain analysis solutions that allow your military or intelligence agency to learn what obstacles you may need to go around, over, or through to reach your desired outcomes.

Proper terrain analysis comes with the ability to present a clear picture of topographical, elevation, and relief parameters. Intergraph provides the solutions your organization needs to meet all of your terrain analysis requirements.

Terrain Analysis Software from Intergraph

Whether being used as part of a military planning system, for humanitarian relief missions, or for environmental engineering planning, terrain analysis software from Intergraph provides an extensive view of the terrain involved as a map or as a 3D interactive model for unparallel degrees of flexibility.

3

Physical Geography and Geomorphology

Physical geographers traditionally have avoided introspective philosophical analysis of their scholarly activities. This situation contrasts sharply with the fierce philosophical debate among human geographers during the past fifteen years. The primary purpose of this chapter is to persuade physical geographers of the virtue of engaging in philosophical introspection; however, it is important to recognise that physical geography is as diverse as human geography. This chapter focuses specifically on the potential contribution of philosophical analysis to geomorphology, but it also informs other sub fields of physical geography, because the issues discussed are central to all branches of the physical sciences. The reason for the neglect of philosophical issues by geomorphologists is unclear. Perhaps it reflects a basic sense of philosophical security. Although the issue of whether geomorphology should be regarded as scientific or historical has been somewhat controversial, much of this discussion has occurred within geology, not geography.

Contemporary geographical geomorphologists appear to be fairly secure in their role as physical scientists, a view that is reinforced by continued emphasis on the dichotomy between human and physical geography. Thus they may be confident that geomorphology is firmly rooted in established philosophical foundations of physics and chemistry. In other words, as physical scientists their philosophical underpinnings are incontrovertible, a perspective strengthened by the limited exposure that many geomorphologists have had to the philosophy of science.

Geomorphologists generally receive little, if any, formal training in philosophical topics at the undergraduate or graduate level. Moreover, in many cases what little training they do receive often comes from human geographers, who simply inform the physical geographers that they are empiricists and then proceed to survey the myriad philosophical perspectives in human geography.

As a result most geographical geomorphologists have perfunctorily embraced either logical positivism (Harvey 1969) or critical rationalism (Haines-Young and Petch 1986). These philosophical perspectives generally are erroneously portrayed as equivalent, except for a minor difference in method: logical positivists attempt to verify hypotheses, but critical rationalists try to falsify them (Johnston 1991, 72).

One purpose of this chapter is to illustrate to human and physical geographers alike that the philosophy of the physical sciences, including geomorphology, is not as secure or uncontroversial as it may seem. Even if the philosophical foundation of the physical sciences is controversial, this situation may not trouble geomorphologists for two reasons. First, they may presume that any debate about philosophical issues concerning the physical sciences should focus on physics or chemistry, not geomorphology, which is derivative from fundamental sciences. This view is misguided because geomorphology as a science deals with distinctive types of complex, unconstrained natural systems that differ from those investigated in laboratory sciences and in many cases uses methods that vary from those employed in laboratory investigations.

It is therefore uncertain that philosophical arguments developed for physics and chemistry directly inform geomorphology. Second, some geomorphologists probably perceive philosophy as an esoteric, unnecessary endeavour that has little relevance for practicing scientists. Although occasional forays into philosophy are necessary for well-rounded scientists, to delve too deeply into such issues often is regarded with suspicion, if not outright contempt, and is sanctioned only if the primary purpose is to seek methodological guidance rather than philosophical insight. This disparaging attitude reflects a basic misunderstanding of the relationship between science and the philosophy of science.

Practicing scientists and philosophers of science have the same goal: to make sense of the world in which they live. Practicing scientists attempt to achieve the goal by interacting directly with the world. Philosophers of science contribute to attainment of the goal by examining the nature of scientific enterprise. Philosophers remind practitioners that they are part of the world, not neutral observers looking in from the outside. Philosophers provide a meta-enquiry into the strengths and weaknesses of science as a cognitive human enterprise directed towards attaining clear, certain, valid knowledge of the world.

This type of analysis provides the foundation for scientific claims to knowledge yet ensures that science as a whole does not become presumptuous about the certainty or truth of this knowledge. Thus any science has an essential relationship with philosophy. Because meaningful debate about the scientific foundation of geomorphology should at least consider, if not be grounded in, existing philosophical frameworks, the following discussion briefly surveys contemporary viewpoints in the philosophy of science, with a specific focus on perspectives pertinent to the physical sciences, and explores their relevance to geomorphology.

Contemporary viewpoints in the philosophy of science can be properly understood only when placed in the appropriate historical context. These viewpoints emerged mainly in response to growing dissatisfaction with logical positivism, or logical empiricism, which dominated the philosophy of science during the first half of the twentieth century (Suppe 1977). Logical positivism represented an attempt to demarcate science from non-science by eliminating metaphysical—that is, unobservable theoretical—elements from scientific theories. The positivist view of science embraced the doctrines of knowledge empiricism, which maintains that all knowledge must be grounded in direct veridical phenomenal experience, and the verifiability theory of meaning, according to which the meaning of a scientific statement is determined by the test conditions under which it can be verified or disconfirmed.

Together these doctrines imply that scientific knowledge is restricted to knowledge of the observable. Knowledge of unobservable, theoretical entities is not possible: for the terms used to describe them to be meaningful, they must be explicitly or partially defined using verifiable, observational language. In other words, references to the unobservable have cognitive significance only if they are defined in terms of the observable. Thus the logical empiricist view of science is committed to metaphysical instrumentalism, or the notion that theoretical terms cannot refer to unobservable properties or entities and that sentences containing these terms do not have truth value (Matheson and Kline 1988).

Although logical empiricists do assign truth value to statements containing theoretical terms, this value is based not on a literal construal of these statements but on a rationally reconstructed construal that purges them of reference to the unobservable. Metaphysical instrumentalism implies that theories are merely logical calculating devices or instruments that allow scientists to derive observational predictions. Logical empiricists asserted that observation provides the foundation for knowledge and that empirical knowledge is accessible independent of theory, that is, theory neutral.

Theory development was viewed as an inductive process involving upward growth from observational facts to theoretical generalisations. Although logical positivists drew a clear distinction between the context of theory discovery and the context of theory justification, they did not directly address the issue of discovery. Instead they focused on the finished product of theorising. Scientific theories were perceived as hierarchical, mathematical constructs with high-level postulates connected deductively to low-level, testable observational statements via correspondence rules, that is, explicit or partial definitions.

Logical empiricists also maintained that scientific progress consists of the accumulation of verified empirical truths and a steady reduction in the number of scientific theories through amalgamation of competitors or absorption of small theories into comprehensive ones. The picture of science initially developed by logical positivists became known as the received view and evolved in a direction which attempted to avoid instrumentalism and to provide theories with a realist interpretation. Those efforts faced tremendous difficulties and rapidly degenerated as various elaborate and largely unsuccessful ad hoc modifications were implemented in an effort to salvage it (Suppe 1977). By the late 1960s many philosophers of science perceived the received view as fatally flawed.

The principal sources of dissatisfaction were its untenable distinction between the observable and the theoretical and its attempt to develop a normative philosophy describing the way in which scientists ought to conduct their business rather than accounting for contemporary and historical scientific activity and scientific claims to knowledge. Between 1960 and 1980 three alternative perspectives emerged as primary competitors in the contemporary philosophy of science: postpositivist empiricism, social constructivism, and scientific realism.

Although these three perspectives differ greatly, they share four commonalities: acknowledgment that the methods of science are theory dependent; increased emphasis on naturalistic philosophical perspectives that fully recognise the constraints imposed on theories of science by the judgements on which scientist base their claims to knowledge and by the natural history of science; elimination of the notion of theory as axiomatised, hierarchical, mathematical constructs and proliferation of less-formalised views of scientific theories; and recognition that no clear distinction exists between the context of discovery and the context of justification.

SOCIAL CONSTRUCTIVISM OR RELATIVISM

Perhaps the strongest blow to logical positivism was the recognition that the methods of science, including observation, are inherently theory laden (Hanson 1958; Kuhn 1970). "Observation is always observation in the light of theories". Because an observation or observation statement can be wrong, it is hypothetical. A theory-neutral form of observation or a significant ontological distinction between the observable and the unobservable is untenable (Maxwell 1962; Putnam 1962). Extreme versions of the theory—ladenness premise provide the basis for relativistic perspectives on scientific knowledge—that is, the epistemological view that the acceptability or unacceptability of scientific claims to knowledge is not objective but depends on social consensus in a specific group of scientists. If observations are strongly dependent on the theories they are used to test, theories are incommensurable because scientists holding different theories will observe different phenomena (Hanson 1958, 4-10), which thereby undermines the adjudicatory power normally accorded to observation.

Moreover, relativists argue that the incommensurability thesis applies not only to modern science but also to the history of science. Based on historical analysis, T. S. Kuhn (1970) and P. K. Feyerabend (1975) concluded that radical conceptual shifts sometimes occur wherein the meaning of scientific terms is not preserved between successive theories or paradigms.

Thus scientists operating under an old theory cannot communicate with those adhering to a new one. During scientific revolutions a complete reconceptualisation of reality occurs that is regulated primarily by social consensus in the scientific community. Science does not progress towards truth; it merely changes its communal perspective on reality. This viewpoint greatly threatens the notion that science reveals truths or theory-independent facts about an objective reality. The aim of science is to solve problems, not to seek the truth.

This relativist perspective did not gain a strong foothold in the United States, but it flourished abroad, specifically at the University of Edinburgh, where it became the foundation of social constructivism (Giere 1988, 50-61). Strong social constructivists (Bloor 1976; Barnes 1982) maintain that theory acceptance and theory change can be explained entirely by social factors. By contrast, weak constructivists (Kuhn 1977; Longino 1990) recognise that other factors have a role in these processes but assert that social factors dominate.

The social-constructivist approach to the study of science is heavily naturalistic, emphasising empirical analysis of actual scientific practice to develop causal explanations of how scientists produce knowledge.

Most of these participant-observer investigations emphasise how social negotiations among scientists play a major role in new claims to knowledge and in the fixation of belief. New scientific knowledge is thus socially constructed in the sense that it is mainly a product of social factors rather than of interaction with the natural world (Collins 1981). Opponents contend that if this assertion be true, the approach is self-refuting because its advocates are using the methods of science to justify the validity of the approach (Niiniluoto 1991a).

POSTPOSITIVIST EMPIRICISM

Although contemporary forms of metaphysical instrumentalism exist (Stegmuller 1976), other postpositivist forms of empiricism reject a strict instrumentalist view of scientific theories (van Fraassen 1980). According to the later perspective, theoretical statements do not acquire meaning only through their connection with the observable, but instead are to be construed literally. They have meaning in and of themselves and are capable of being true or false.

By accepting that statements about unobservable phenomena have truth value, this type of empiricism avoids the dilemma posed by the lack of a clear ontological distinction between the observable and the unobservable. However, it does maintain that directly observable phenomena have privileged epistemological status.

In other words, knowledge about directly observable phenomena has a more certain evidential basis than do claims to knowledge about unobservable phenomena. This type of empiricism has been characterised as epistemological instrumentalism.

Although epistemological instrumentalism acknowledges that the theoretical entities in theories are best thought of as potentially real, it is still antirealist in that it maintains that "good" scientific theories do not have to be true; thus acceptance of a theory does not mandate belief in its truth or in the reality of the unobservable entities it postulates.

In other words, the truth value of a theory is not the basis for theory choice. Instead, the aim of a science is to generate theories that are empirically adequate (van Fraassen 1980) or that solve problems (Laudan 1977, 1984). Acceptance of a theory commits scientists only to the belief that the theory is empirically adequate or solves problems effectively, not to the position that it is true. Empirical adequacy refers to a high degree of conformity between directly observable phenomena and what a theory predicts about these observable aspects of the world.

Epistemological instrumentalism is strongly empiricist in that it involves a search for truth only about observable phenomena and maintains that science provides no means to justify belief in theoretical entities. According to the contemporary empiricists, ultimately the only basis for evaluating theoretical hypotheses is observational evidence. Because all theoretical statements extend beyond this observational evidence, they are underdetermined in the sense that they are less probable than the empirical evidence. Observational evidence cannot determine the truth of theoretical propositions, and belief in the truth of such statements always involves greater epistemic risk than does belief in the truth of statements about the observable (van Fraassen 1985). Although this form of empiricism assigns privileged epistemological status to statements about observable phenomena, it fails to provide a definitive philosophical argument to support this status.

SCIENTIFIC REALISM

Contemporary scientific realism derives from philosophical ideas espoused by Charles S. Peirce near the end of the nineteenth century and by Karl Popper since 1930 (Niiniluoto 1991b). Peirce and Popper both interpreted scientific knowledge as corrigible and rejected the received-view notion of naive realism that science progresses through the steady accumulation of verified truths about the observable world. Nevertheless, both argued that truth plays an important role in scientific progress because over time, according to Peirce, science moves towards the truth about reality or, according to Popper, develops theories that are more and more truthlike. Popper (1968, 423-425) also denied metaphysical instrumentalism and embraced a realist interpretation of scientific theories.

Numerous versions of realism exist in contemporary philosophy of science. These variations can be divided into three categories: internal realism, entity realism, and critical realism (Niiniluoto 1991b). Internal realists reject the view of truth as correspondence between statements about the world and facts about an objective, accessible reality. Instead, they replace this definition with a pragmatic, epistemic notion of truth, such as ideal acceptability or eventual perfection of scientific explanations (Newton-Smith 1989).

Entity realists subscribe to the correspondence theory of truth, but they accept existence claims only about unobservable entities that can be manipulated experimentally to produce detectable effects, such as electrons, neutrinos, or positrons, and deny a realist interpretation of theoretical laws.

Although these realist philosophical positions have merit, the label scientific realism is normally associated with critical realism, which most commonly serves as the focal point for opposing philosophical criticism by social constructivists and postpositivist empiricists. Critical realists maintain four assumptions. Terms in scientific theories that describe unobservable entities or causal mechanisms should be interpreted realistically: the terms refer to phenomena that genuinely exist. Reality exists independently of human thought, and the description of this reality embodied in scientific theories is largely independent of theoretical commitments. The methods of science provide a means for determining the approximate truth of all aspects of scientific theories, including statements about unobservables.

The historical progress of science consists of the development of successive theories that become more truthlike over time, with later theories building on the knowledge embodied in previous ones; thus current scientific theories in mature sciences are at least approximately true (Boyd 1983; Niiniluoto 1984). This perspective is a strong version of realism in that it includes ontological and epistemological components. Clearly, critical realism adopts the perspective that the aim of science is to seek the truth, not merely to solve problems. Critical realists readily acknowledge that all aspects of scientific enquiry are theory laden. They maintain that this characteristic of science permits knowledge of a theory-independent world and accounts for the reliability of this knowledge. This assertion is supported by an abductive argument, or inference to the best explanation, which has the following structure.

- Premise 1—If scientific theories are approximately true, they will be empirically successful.
- Premise 2—Many scientific theories are empirically successful.
- Probable conclusion—Many scientific theories are approximately true.

In other words, realists argue that the only plausible explanation for the overwhelming empirical success of a heavily theory-dependent enterprise like science is that the theories on which it is based are at least approximately true. This line of reasoning is known as the no-miracles argument: if the best current theories are not at least approximately true, the empirical success of modern science must be viewed as a miracle.

According to critical realists, current theories are approximately true and are the foundation for scientific progress. Although an infinite number of possible explanations exists for any anomalous empirical result, scientists do not assign equal weight to each explanation. Instead, based on their current theoretical knowledge, they immediately discount all but a few: existing theory plays an evidential role in choosing between competing theories and serves as a basis for the development of new theories (Boyd 1983). Theoretical knowledge is evidentially as important as, if not more important than, empirical knowledge. Also, the theory ladenness of observation is not problematic because it is the true or approximately true theoretical knowledge on which scientific observational procedures are based that allows access to new theoretical knowledge about theory-independent reality.

This situation is best exemplified when scientists use theory to develop sophisticated measurement techniques or instruments that permit detection of the effects of unobservable entities and causal mechanisms (Brown 1990). The epistemological component of critical realism requires justification in light of the constructivist argument that science occasionally undergoes radical conceptual shifts in which previous theories are completely abandoned, with resulting semantic incommensurability between scientific theories.

The notion of radical conceptual shifts greatly threatens the realist image of scientific progress. For current theories to be approximately true, they must preserve as limiting cases antecedent theories that were also approximately true. However, if successive theories are truly incommensurable, this conception of scientific progress is difficult to defend. To address this problem, realists adopt a causal or referential theory of meaning wherein the meaning of a theoretical term that describes an unobservable entity is determined not by the properties of the entity, which depend on the theory, but by an introducing event that puts scientists in the appropriate causal relationship with the entity (Putnam 1973). Usually this appropriate relationship involves a situation wherein the entity is present and produces certain effects. The properties assigned to the entity may differ between theories or over time as theories change, but as long as scientists agree that the entity responsible for certain effects under theory A is the same entity responsible for effects under theory B, reference and meaning are preserved. The development of true definitions of scientific phenomena based on assignation of properties is an a posteriori theoretical issue rather than an a priori problem of convention (Boyd 1990).

Critical realism also rejects metaphysical realism, the notion that reality consists of facts that exist independently of human conceptualisation and that one true theory is capable of describing this reality. Instead, critical realists acknowledge that facts are always created by humans in the sense that they are in part linguistic entities that presuppose the existence of human minds and conceptual frameworks (Niiniluoto 1984, 177-178). However, in contrast to constructivists, critical realists maintain that even though language is required to carve up the world into facts that become the object of scientific investigations, the truth of statements about these facts is determined not by the conceptual framework but by the causal interaction with an objective reality. This view of science is fundamentally nonreductionist because it implies that no single language or science has privileged status concerning access to reality and that unrelated conceptual frameworks can reveal different truths about reality (Boyd 1989).

LANDFORMS OF GEOMORPHOLOGY

Landform

A landform or physical feature in the earth sciences and geology sub-fields, comprises a geomorphological unit, and is largely defined by its surface form and location in the landscape, as part of the terrain, and as such, is typically an element of topography. Landform elements also include seascape and oceanic waterbody interface features such as bays, peninsulas, seas and so forth, including sub-aqueous terrain features such as submersed mountain ranges, volcanoes, and the great ocean basins.

Physical Characteristics

Landforms are categorised by characteristic physical attributes such as elevation, slope, orientation, stratification, rock exposure, and soil type.

Gross *physical features or landforms* include intuitive elements such as berms, mounds, hills, ridges, cliffs, valleys, rivers, peninsulas and numerous other structural and size-scaled (i.e. ponds vs. lakes, hills vs. mountains) elements including various kinds of inland and oceanic waterbodies and sub-surface features.

Hierarchy of Classes

Oceans and continents exemplify the highest-order landforms. Landform elements are parts of a high-order landforms that can be further identified and systematically given a cohesive definition such as hill-tops, shoulders, saddles, foreslopes and backslopes.

Some generic landform elements including: pits, peaks, channels, ridges, passes, pools and plains, may be extracted from a digital elevation model using some automated techniques where the data has been gathered by modern satellites and stereoscopic aerial surveillance cameras. Until recently, compiling the data found in such data sets required time consuming and expensive techniques of many man-hours.

Terrain (or *relief*) is the third or vertical dimension of *land surface*. Topography is the study of terrain, although the word is often used as a synonym for relief itself. When relief is described underwater, the term bathymetry is used. In cartography, many different techniques are used to describe relief, including contour lines and TIN (Triangulated irregular network).

Elementary landforms (segments, facets, relief units) are the smallest homogeneous divisions of the land surface, at the given scale/resolution. These are areas with relatively homogenous morphometric properties, bounded by lines of discontinuity. A plateau or a hill can be observed at various scales ranging from few hundred meters to hundreds of kilometers. Hence, the spatial distribution of landforms is often scale-dependent as is the case for soils and geological strata.

A number of factors, ranging from plate tectonics to erosion and deposition, can generate and affect landforms. Biological factors can also influence landforms— for example, note the role of vegetation in the development of dune systems and salt marshes, and the work of corals and algae in the formation of coral reefs. Landforms do not include man-made features, such as canals, ports and many harbors; and geographic features, such as deserts, forests, grasslands, and impact craters.

Many of the terms are not restricted to refer to features of the planet Earth, and can be used to describe surface features of other planets and similar objects in the Universe. Examples are mountains, polar caps, and valleys, which are found on all of the terrestrial planets.

Landform Construction and Destruction

Megageomorphology: In keeping with the regional scale of our examination of major geomorphic provinces we will begin this course by considering tectonic landscapes on the scale of whole mountain ranges and some of the general processes that shape them.

Geomorphologists have also traditionally divided landforms into constructional and destructional categories.

Constructional landforms are those that have been or are being built (increasing in mass, height, or area). Examples include tectonic forms such as fault scarps, fault block mountains and valleys (horsts and grabens), rift valleys, monoclines, salt domes, and a variety of volcano landforms. In general, constructional landforms are more common in regions of active tectonism, usually associated with plate margins. Destructional landforms are those that are decreasing in mass, height, or area over time and are primarily the products of erosion and weathering. In truth, all landforms are in part destructional because erosion affects all landforms that lie above sea level on the Earth's surface. However, many landforms are primarily the result of such processes.

More recently, geomorphologists have come to view all landscapes as the products of some combination of three factors:

- tectonics (orogeny, epeirogeny, extension, volcanism)
- erosion
- climate.

None of these factors operates independently-all are in some way functions of the others:

- Tectonics are clearly affected by erosion-removal of crustal mass by erosion leads to isostatic rebound and uplift. In general, about 80% of the height of a landmass removed by erosion will be replaced from below by isostatic readjustment. Some geologists refer to the "pull of erosion" to describe this phenomenon.
- Uplift accelerates erosion, which, as a general rule, increases in intensity with increasing elevation. High elevations also host less vegetation, which armors the land against erosion at low elevations.
- Erosion is a function of climate in many ways. Humid climates as a rule erode faster than arid climates and the effects of erosion on the landscape differ. Humid climates tend to result in the formation of deeply incised valleys. Arid regions erode more evenly. Cold climates result in mountain glaciers that can carve extremely steep topographic features. In fact, mountain glaciers may be Earth's most effective agents of erosion. Extremely frigid climates host large ice sheets that do very little erosion because they are frozen to the bedrock on which they rest.

- Climate itself is greatly affected by tectonism. High landmasses change the balance of heating on the surface and block or redirect the flow of weather systems. As a landmass is elevated its climate cools and may become more humid. This in turn can accelerate erosion. Windward of a mountain range the climate is usually humid as ascending air masses cool and drop their moisture. Leeward of the range is a rain shadow and very arid conditions.

Isostatic mountain building: Climate and erosion can also work together to create uplifted mountain peaks. An arid plateau will support few rivers and tend to erode slowly and uniformly. As material is removed across the plateau isostatic readjustment will cause some uplift and the average land surface will slowly be lowered. If however, the plateau is in moist climate with abundant rivers or glaciers, then abundant, deep valleys can form, removing a large mass of crust, but leaving some crust standing as peaks. Isostatic readjustment will elevate the plateau so that its average mass will be lower than before erosion, but this can result in the peaks being elevated to a higher level than before.

The Three Stages of Mountains: It is probably incorrect to say the mountains are uplifted and then eroded back to sea level. Rather, isostacy, climate and erosion interact to create three stages in the life of a mountain range.

1. Formative Stage-a tectonic event thickens or heats the crust and creates uplift with rates that exceed erosion rates. As the topography rises erosion rates increase.
2. Steady State-erosion rates rise to match uplift rates or uplift slows to match erosion. A steady state results with uplift and erosion rates balanced. Because uplift rates and erosion rates create a negative feedback loop (increasing uplift raises the mountains which intensifies erosion and lowers the mountains) the equilibrium elevation of the mountains may remain stable for millions of years.
3. Decline-if uplift diminishes sufficiently erosion may come to dominate and slowly lower the elevation of the range. This is a slow process because isostatic rebound will replace much of the elevation lost to erosion. By the time a mountain range is worn back to the average level of continental crust, a thickness of crust several times greater than the height of the mountains at their highest can be removed by erosion. For example, it is estimated from mineral studies that the metamorphic rocks of New York City and Connecticut were once buried under about 7 miles of overlying rock. This is not to say that the mountains above them were 7 miles high, only that this thickness of crust has been pulled upward by isostatic readjustment over the last 450 million years.

Major Geomorpho-tectonic Regions

- Coastal Plains-these are extensive regions of flat-lying to gently seaward dipping sediments that build up as a passive continental margin subsides over millions of years. They are areas of very low relief. Generally, coastal plains are absent in tectonically active areas because active uplift of the coastline prohibits their development.
- Orogenic Belts-these are elongate regions that form from collisional tectonics along active continental margins. In presently active orogenic belts the metamorphic rocks may not yet be exposed on the surface-only after extensive erosion of the overlying, intensely deformed sedimentary cover are the metamorphic roots of the mountains exposed. Where they are generated by subduction of oceanic crust beneath the continent they are highly intruded by magmatic plutons.
- Fold and Thrust Belts-marginal to orogenic belts are regions where the sedimentary cover has been folded and thrust faulted by compressional stresses without metamorphism. Low angle thrust faults create blocks of sedimentary crust stacked one on another, suggesting large amounts of shortening of the upper crust (often refered to as "thin-skined tectonics"). Erosion of ancient fold and thrust belts near sea level produces a distinctive ridge and valley topography with surface expressions of underlying syncline and anticline folds.
- Plateaus-these are tectonically elevated regions of undeformed sedimentary rock (the Tibetan Plateau and the Chilean Altiplano are a somewhat different type of plateau-elevated regions within an orogenic belt created by slow erosion due to rain shadow climates). Plateaus developed in in humid regions can be very mountainous due to stream dissection. An extreme form of stream dissection is seen in the canyonlands of the Colorado Plateau, where rivers such as the Colorado have carved mile-deep canyons such as the Grand Canyon.
- Stable Interior/Shield-the interior regions of continents are generally low relief areas with very mild stream dissection. Shield areas in the center of a continent are underlain by ancient, deeply eroded metamorphic and metaigneous rocks. Surrounding the shield regions are areas of the stable interior covered with a thin veneer of nearly horizontal sedimentary strata. These strata are very subtly deformed into broad structurally upwarped regions called domes and downwarped regions called basins which generally do not have any topographic expression.
- Extensional Regions-regions of high mantle heat flow beneath continental crust can cause broad crustal doming and the formation of

fault block mountains bounded by grabens and half grabens. Large grabens are also known as rift valleys when they are associated with continental rifting. These features form as the crust is thinned and stretched, creating tensional forces and normal faults. The extensive Basin and Range region of Utah and Nevada in the western US has been generated in the last 17 million years by extension of the western US by as much as 100 miles in the east-west direction. Extensional tectonics are often accompanied by igneous intrusives and surface eruptions as magma works its way upward along faults in the thinned crust.

MORPHOMETRIC ANALYSIS

Formulas for right circular cones have been fitted to the configurations of alluvial fans, logarithmic spirals have been used to describe certain shapes of beaches, and drumlins, spoon-shaped glacial landforms, have been found to accord to the form of the lemniscate curve. With regard to drainage basins, many quantitative measures have been developed to describe valley side and channel slopes, relief, area, drainage network type and extent, and other variables. Attempts to correlate statistically parameters defining drainage basin characteristics and basin hydrology, as in studies of sediment yield, are generally designated as morphometric analyses.

Morphometrics

Morphometrics Morphometrics refers to the quantitative analysis of *form*, a concept that encompasses size and shape. Morphometric analyses are commonly performed on organisms, and are useful in analyzing their fossil record, the impact of mutations on shape, developmental changes in form, covariances between ecological factors and shape, as well for estimating quantitative-genetic parameters of shape. Morphometrics can be used to quantify a trait of evolutionary significance, and by detecting changes in the shape, deduce something of their ontogeny, function or evolutionary relationships. A major objective of morphometrics is to statistically test hypotheses about the factors that affect shape. "Morphometrics", in the broader sense of the term, is also used to precisely locate certain areas of featureless organs such as the brain, and is used in describing the shapes of other things.

Forms of Morphometrics

Three general approaches to form are usually distinguished: traditional morphometrics, landmark-based morphometrics and outline-based morphometrics.

"Traditional" Morphometrics

Traditional morphometrics analyzes lengths, widths, masses, angles, ratios and areas (image). In general, traditional morphometric data are measurements of size. A drawback of using many measurements of size is that most will be highly correlated; as a result, there are few independent variables despite the many measurements. For instance, tibia length will vary femur length and also with humerus and ulna length and even with measurements of the head. Traditional morphometric data are nonetheless useful when either absolute or relative sizes are of particular interest, such as in studies of growth. These data are also useful when size measurements are of theoretical importance such as body mass and limb cross-sectional area and length in studies of functional morphology. However, these measurements have one important limitation: they contain little information about the spatial distribution of shape changes across the organism.

Landmark-based Geometric Morphometrics

In landmark-based geometric morphometrics that spatial information is contained in the data because the data are coordinates of "landmarks": discrete anatomical loci that are arguably homologous in all individuals in the analysis. For example, where sutures intersect is a landmark as are intersections between veins on an insect wing or leaf. So are foramina, small holes through which veins and blood vessels pass. These points can be regarded as the "same" point in all specimens in the study. Landmark-based studies have traditionally analyzed 2D data, but with the increasing availability of 3D imaging techniques, 3D analysis are becoming more feasible even for small structures such as teeth. Finding enough landmarks to provide a comprehensive description of shape can be difficult when working with fossils or easily damaged specimens. That is because all landmarks must be present in all specimens, although coordinates of missing landmarks can be estimated. The data for each individual consists of a *configuration* of landmarks.

There are three recognized categories of landmarks. Type 1 landmarks are locally defined, meaning that they are defined in terms of structures close to that point; for example, an intersection between three sutures, or intersections between veins on an insect wing are locally defined and surrounded by tissue on all sides. Type 3 landmarks, in contrast, are defined in terms of points far away from the landmark, and are often defined in terms of a point "furthest away" from another point. Type 2 landmarks are intermediate; this category includes points such as the tip structure, or local minima and maxima of curvature. They are defined in terms of local features, but they are not surrounded on all sides. In addition to landmarks, there are "semilandmarks," points along a curve. Their position along the curve is arbitrary but these points provide information about curvature in two or three dimensions.

CONTEMPORARY PERSPECTIVES AND GEOMORPHOLOGY

Geomorphologists will develop better understanding of the scientific foundations of their field through philosophical analysis. Although philosophical self-examination could be conducted in an independent manner, it is best pursued in the context of ongoing debate in the philosophy of the physical sciences. This type of analysis allows geomorphologists to address questions such as: what constitutes a scientifically valid explanation; are there different types of acceptable explanations, and, if so, what relationships, if any, exist among them; what is a geomorphological theory; to what extent is geomorphological knowledge a product of social, cultural, political, and ethical factors; are geomorphologists problem solvers or truth seekers; how do geomorphologists justify and defend their claims to knowledge; are these methods consistent with the aims of the field; and how are responses to these questions similar to or different from those provided by other scientific disciplines?

Although social constructivism, postpositivist empiricism, and scientific realism are the primary focuses of current philosophical debate, other viewpoints also may contain ideas relevant to geomorphology. Specifically, it may be worthwhile to explore philosophical perspectives that have emerged in disciplines such as biology and physics. However, there is no reason to presuppose that a philosophical framework for geomorphology will be merely a restatement of the philosophy of another discipline. Because geomorphology is concerned with distinctive types of natural systems that include synergistic physical and biological elements and employs characteristic investigative methods, it cannot be reduced to the underpinning disciplines.

One predominant idea in the philosophy of science merits serious consideration by geomorphologists. The three contemporary philosophical perspectives discussed above, though widely divergent in many respects, uniformly subscribe to the theory-laden view of scientific observation. Given their many differences, this commonality suggests something of uncommon significance. It also contrasts sharply with the theory-neutral interpretation of observation that appears to be a de facto—albeit often unconscious—belief of many geomorphologists. Philosophers have devoted considerable attention to the topic of scientific observation. Although a universally accepted definition has not emerged from this analysis, most philosophers agree that scientific observation cannot be equated only with perception; instead, it is a complex process that includes identification, interpretation, and description. The relationship between theory and observation in geomorphology deserves more consideration than it has received.

The philosophy of science not only provides a means of evaluating contemporary geomorphology but also serves as a framework for assessing the historical development of the discipline. Most available philosophical perspectives incorporate the explicit caveat that they apply to mature sciences only. Even scientific realists acknowledge that scientific theories may be largely social constructs in an inchoate science; therefore, it becomes difficult to assess whether a scientific discipline corresponds to the realist conception of scientific progress until it has developed a substantial history. Whether or not geomorphology qualifies as a mature science is unclear; nevertheless, any tentative assessment or debate about the nature of such an assessment should occur in the context of existing philosophical ideas.

In geomorphology the transition from the Davisian view of landscape development to the modern emphasis on systematic process-oriented investigations has been explicitly and implicitly characterised as a Kuhnian-style conceptual change or paradigm shift. On the assumption that the majority of geomorphologists prefer to view their field from a realist perspective, this characterisation of the historical development of geomorphology is inconsistent with the prevailing philosophical perspective, because the Kuhnian theory of science incorporates the notion that science does not progress towards the truth. The challenge for realists is to show how many theoretical constructs embodied in the Davisian view of geomorphology, including references to unobservables, have been preserved in contemporary geomorphic theories (Rhoads forthcoming). On the other hand, process geomorphologists must recognise that adopting a realist perspective does not necessitate that a truly scientific approach implies that all geomorphological problems must be described in the language of physics. Contemporary scientific realism explicitly acknowledges that no scientific discipline has privileged status with regard to the truth.

Exploration of philosophical issues in geomorphology should also enhance disciplinary unity. Contention in contemporary geomorphology centres on differences in regulative principles, types of scientific arguments, and characteristics of theory employed by scientists, all of whom consider themselves geomorphologists (Rhoads and Thorn 1993). Although the point of contention is scientific methodology, the contention itself is clearly philosophical in nature. In other words, it is not possible to resolve conflicts between competing methodologies within science itself; instead, resolution of these differences must occur within philosophy (Montgomery 1991). As this issue and others like it are examined philosophically, differences in opinion will emerge about which viewpoint provides the clearest perspective on geomorphology. However, here the goals of science and those of philosophy differ. Whereas practicing scientists define the success of their endeavours in empirical terms, such as confirmation and acceptance of factual knowledge, philosophers deal with concepts and define success quite differently.

Philosophers commonly exhibit great respect for the intellectual contributions of people who hold ideas directly opposed to their own. Success in philosophy is measured not by declaring that one perspective is right and that others are wrong, as tends to happen when methodological issues are debated in science, but by the degree to which opposing viewpoints have reached a consensus as to which issues are central to a specific debate and by the level of rational sophistication of the arguments. Philosophical insight by necessity depends on intellectual disagreement and debate. Viewed in this light, philosophical introspection provides an excellent antidote to scientists who wish to divide their colleagues into winners and losers on the basis of methodological preferences. Indeed, it offers the opportunity to embrace methodological diversity in a substantive and constructive manner and to enhance the intellectual depth of the discipline.

PHYSICAL CONDITIONS OF THE EARTH'S INTERIOR

Three centuries ago, the English scientist Isaac Newton calculated, from his studies of planets and the force of gravity, that the average density of the Earth is twice that of surface rocks and therefore that the Earth's interior must be composed of much denser material. Our knowledge of what's inside the Earth has improved immensely since Newton's time, but his estimate of the density remains essentially unchanged. Our current information comes from studies of the paths and characteristics of earthquake waves travelling through the Earth, as well as from laboratory experiments on surface minerals and rocks at high pressure and temperature. Other important data on the Earth's interior come from geological observation of surface rocks and studies of the Earth's motions in the Solar System, its gravity and magnetic fields, and the flow of heat from inside the Earth.

The planet Earth is made up of three main shells: the very thin, brittle crust, the mantle, and the core; the mantle and core are each divided into two parts. All parts are drawn to scale on the cover of this publication, and a table at the end lists the thicknesses of the parts. Although the core and mantle are about equal in thickness, the core actually forms only 15 per cent of the Earth's volume, whereas the mantle occupies 84 per cent. The crust makes up the remaining 1 per cent. Our knowledge of the layering and chemical composition of the Earth is steadily being improved by earth scientists doing laboratory experiments on rocks at high pressure and analyzing earthquake records on computers.

The Crust

Because the crust is accessible to us, its geology has been extensively studied, and therefore much more information is known about its structure and composition than about the structure and composition of the mantle and core. Within the crust, intricate patterns are created when rocks are redistributed and deposited in layers through the geologic processes of eruption and intrusion of lava, erosion, and consolidation of rock particles, and solidification and recrystallization of porous rock.

By the large-scale process of plate tectonics, about twelve plates, which contain combinations of continents and ocean basins, have moved around on the Earth's surface through much of geologic time. The edges of the plates are marked by concentrations of earthquakes and volcanoes. Collisions of plates can produce mountains like the Himalayas, the tallest range in the world. The plates include the crust and part of the upper mantle, and they move over a hot, yielding upper mantle zone at very slow rates of a few centimeters per year, slower than the rate at which fingernails grow. The crust is much thinner under the oceans than under continents. The boundary between the crust and mantle is called the Mohorovicic discontinuity; it is named in humour of the man who discovered it, the Croatian scientist Andrija Mohorovicic. No one has ever seen this boundary, but it can be detected by a sharp increase downward in the speed of earthquake waves there. The explanation for the increase at the Moho is presumed to be a change in rock types. Drill holes to penetrate the Moho have been proposed, and a Soviet hole on the Kola Peninsula has been drilled to a depth of 12 kilometers, but drilling expense increases enormously with depth, and Moho penetration is not likely very soon.

The Mantle

Our knowledge of the upper mantle, including the tectonic plates, is derived from analyses of earthquake waves; heat flow, magnetic, and gravity studies; and laboratory experiments on rocks and minerals. Between 100 and 200 kilometers below the Earth's surface, the temperature of the rock is near the melting point; molten rock erupted by some volcanoes originates in this region of the mantle. This zone of extremely yielding rock has a slightly lower velocity of earthquake waves and is presumed to be the layer on which the tectonic plates ride. This low-velocity zone is a transition zone in the upper mantle; it contains two discontinuities caused by changes from less dense to more dense minerals. The chemical composition and crystal forms of these minerals have been identified by laboratory experiments at high pressure and temperature. The lower mantle, below the transition zone, is made up of relatively simple iron and magnesium silicate minerals, which change gradually with depth to very dense forms. Going from mantle to core, there is a marked decrease in earthquake wave velocity and a marked increase in density.

The Core

The paths curve because the different rock types found at different depths change the speed at which the waves travel. Solid lines marked P are compressional waves; dashed lines marked S are shear waves. S waves do not travel through the core but may be converted to compressional waves on entering the core. Waves may be reflected at the surface. The core was the first internal structural element to be identified. It was discovered in 1906 by R.D. Oldham, from his study of earthquake records, and it helped to explain Newton's calculation of the Earth's density. The outer core is presumed to be liquid because it does not transmit shear (S) waves and because the velocity of compressional (P) waves that pass through it is sharply reduced. The inner core is considered to be solid because of the behaviour of P and S waves passing through it. Cross section of the whole Earth, showing the complexity of paths of earthquake waves. The paths curve because the different rock types found at different depths change the speed at which the waves travel. Solid lines marked P are compressional waves; dashed lines marked S are shear waves. S waves do not travel through the core but may be converted to compressional waves on entering the core. Waves may be reflected at the surface.

Data from earthquake waves, rotations and inertia of the whole Earth, magnetic-field dynamo theory, and laboratory experiments on melting and alloying of iron all contribute to the identification of the composition of the inner and outer core. The core is presumed to be composed principally of iron, with about 10 per cent alloy of oxygen or sulfur or nickel, or perhaps some combination of these three elements.

The Structure of the Moon

The Moon, our fellow-traveller in space, has a diameter half that of the Earth's core, and it revolves around the Earth, as all the planets revolve around the Sun, under the force of gravity. Moonquakes of very low energy are caused by land tides produced by the pull of Earth's gravity, and, from analysis of moonquake data, scientists believe the Moon has two layers: a crust, from the surface to 65 kilometers depth, and an inner, more dense mantle from the crust to the center at 3,700 kilometers.

The crust is presumed to be com-posed primarily of rocks containing feldspar, calcium aluminum silicate, and lesser pyrox-ene, iron and magnesium silicate; the crust also contains basalt in the mares, which con-tains less iron and more titanium than earth basalt. The mantle is thought to be made up of calcic peridotite, containing both pyroxene and feldspar.

4

Coastal Geomorphology

Nova Scotia has a great variety of coastal landforms. Erosional features are predominant, but the landforms produced vary just as to the geology and glacial history. Depositional landforms, beaches and marshes, are being formed as a result of erosion and transportation of unconsolidated material. But many of these deposits are, in turn, being eroded and submerged by the rising sea level. These descriptions of coastal landforms should be read in conjunction with the habitat descriptions.

Ancient Shorelines

Parts of Nova Scotia's coastline are remnants of ancient shorelines formed as sea level changed during and since the last glaciation. Some of these features are exposed, while others are now under water; they provide clues to the development of the province over the last 10,000 years or more.

In the Northumberland Strait, for instance, a series of submerged terraces cut into the less-resistant rocks document successive sea levels of the post-glacial period.

In the western and northern extremities of Nova Scotia, an elevated shoreline is found. This terrace, or bench, cut in the bedrock is a distinctive landscape feature in St. Marys Bay and on the east and west of northern Cape Breton Island.

In St. Marys Bay it is assumed to be 125,000 years old and to have been formed in a warmer, interglacial period. This age is assumed because extensively weathered glacial tills are found on the bench. Elevation of this ancient bench above present sea level is only slight, averaging 4–6 m. Gravel beds, sea stacks and striated potholes have also been found on the surface.

By inference, the ancient wave-cut terrace north of St. Margaret Village which is at a similar elevation to the bench, is the same age. Remnants of what is probably the same platform are found around Cape George, Antigonish County.

Marine Limit

The maximum inland penetration of the post-glacial sea level is marked by relict coastal features similar to those now found at the shoreline. These features are confined to the north and west of a line joining Yarmouth, Windsor, Truro and Northport. The marine limit is highest on Digby Neck and at Brier Island. Well-preserved beaches are found at 36 m above mean sea level. At Sandy Cove a wave-cut cliff exposes 40 m of stratified sand and gravel, the top of which has been levelled by wave erosion. To the south and east the limit declines in altitude.

A record of the Acadian Glaciation. Vlarine gravel: deposited in immediate pose-glacial time, higher sea level.

- *Light-Grey Slate Till*: Locally derived till, indicating early and late local ice caps.
- *Debris*:Angular fragments from periglacial conditions prior to ice advance.
- *Marine Gravel(rusty)*: Deposited prior to last glacial advance.
- *Marine Platform*: Fossil wave-cut platfrom just above present sea level, backed by fossil cliff, 120,000 years ago and earlier.

A delta on the Sissiboo River at Weymouth is 25 m high. At Smith Cove in the Annapolis Valley stratified materials reach 33 m. Raised beaches near Woodside are 25 m high and 5.5 km inland from the present coast. At the same times as these beaches were being formed, low-lying areas of the modern Annapolis Valley were submerged. Marine clays were deposited in these submarine basins. Near Bridgetown and Lawrencetown, these clays are now exposed and have been mapped as Lawrencetown and Fash soils in Annapolis County.

On the Cobequid shore, a series of conspicuous raised beaches reaches a maximum height of 40 m above present sea level at Advocate where a replica of the present harbour deposits is found on the hillside behind the village. In the same District, at Parrsboro, the raised beaches are 22 m high. The marine limit declines to present sea level at Truro. It is assumed that south and east of the line joining Yarmouth, Truro and Northport, post-glacial sea levels have never been higher than they are at present.

This tilting of the land mass is evidence for a greater glacial depression of the land in the Bay of Fundy area than in eastern Nova Scotia.

Buried Valleys

Lower sea levels from 15,000 to 8000 years ago, and probably from earlier periods, caused the rivers to extend well beyond the present estuaries. In some cases where the rivers flow over easily eroded rocks, especially the Triassic sandstones of the Bay of Fundy, the river valleys were cut down deeply into the bedrock. Buried valleys in the bedrock have been found at Truro and Brookfield and in the Annapolis Valley. Both the North and Salmon rivers at Truro have similar buried valleys, extending 6 km and 8 km respectively under the present floodplain north of Truro. In places, the channel of the Salmon River reaches 40 m below the present land surface. Buried channels in the Truro and Kentville areas are filled with glacial outwash sands and gravels.

Submergence

Sometime before 8000 years ago in the Bay of Fundy, and before 15,000 years ago elsewhere in Nova Scotia, the sea once again began to encroach upon the land. Forests had developed on the exposed sea floor, which was probably a maximum of 37 m below present sea levels in the Bay of Fundy.

The submerged forests to be found on the Atlantic and Bay of Fundy coasts are probably the most dramatic evidence of sea-level change. Pine and Beech stumps more than a metre in diameter occur at these sites and are well rooted in Podzol soils resembling those of the surrounding uplands. On the soil surface a forest-floor humus of partly decomposed leaves, twigs and cones can still be found. Many scattered examples of these forests occur along the Atlantic Coast, extending in places well beyond the low-tide level. Well-known examples are Yarmouth Harbour, Sunday Point, Cape Sable Island, Lunenburg, McNabs Island in Halifax Harbour and Bon Portage in Yarmouth County. Most of these stumps are between 5000 and 3000 years old. It has been deduced that their submergence was due largely to recent crustal subsidence, which over he last 4000 years has averaged 30 cm per century.

Rocky Shores

Where resistant rocks occur, there are insufficient unconsolidated deposits to build beaches and marshes. The only significant physical process in such circumstances is the hydraulic "quarrying" of large blocks by storm waves, any glacial till overburden having been simply swept off the shore by large waves. Two types of rocky shores without cliffs can be recognized: granitic rocks and Meguma rocks, principally quartzite. Both are part of the gently seaward dipping surface of the Atlantic upland. Only where less-resistant volcanic rocks occur at Louisbourg and Fourchu are sea cliffs formed.

The granite headlands of the Atlantic Coast have changed little despite their submergence and exposure to waves. The thin cover of glacial till, however, has been removed to expose the glacially smoothed bedrock. The gradual seaward slope of the land continues under the ocean without interruption. Where the granite is more heavily jointed, waves have been more effective in excavating large blocks. In the quartzites of the Meguma rocks, joints are more closely spaced. As a result "quarrying" has been more effective. Modification of these rocks has not been sufficient to produce cliffs, but tide pools are more numerous, especially where the rocks outcrop parallel to the coast, as on much of the Eastern Shore.

Cliffs

Cliffs are being formed in less-resistant rocks, such as sandstones, shales and basalt, and in unconsolidated materials, especially drumlins. Ancient cliffs, especially in northern Nova Scotia, were formed over long periods before the last ice advances and often relate to structural weaknesses, such as fault lines. This feature can be attributed to the general tilting of the province to the southeast. Notable ancient cliffs are found in Cape Breton north of the Margaree Valley on the west coast. On the mainland, Cape George is a southerly extension of this same geological structure. Northern limits of the southern upland Meguma rocks form seacliffs on the south side of Chedabucto Bay which is a fault-line scarp, and the south side of St. Mary's Bay, where the strike parallels the coast.

The most rapidly eroding cliffs are those in partially submerged drumlins on the Atlantic Coast and in unresistant horizontal sandstones of the Bay of Fundy, Northumberland Strait and the Sydney Coalfield.

Mean rates for the Bay of Fundy are about 0.5 m/year and for exposed Atlantic Coast drumlins are about 1.0 m/year. Eroded drumlins are marked by a lag deposit, just below low-tide level, of the larger boulders. All other material has been winnowed from the drumlin and redistributed. As long ago as 1929, Johnson attempted to recreate a map of the former drumlin field in Mahone Bay from hydrographic charts of shoals. He mapped nearly sixty shoals or lag deposits which must have been drumlins within the past 10,000 years, and there are many other bars and small islands which may at one time have been drumlins.

Observations of a drumlin recently exposed to wave attack at Chezzetcook Inlet recorded rates of erosion averaging 5.4 m/year. In contrast, the erosion of bedrock cliffs in the Bay of Fundy, Northumberland Strait and Cabot Strait creates a wave-cut platform in the bedrock as the cliff recedes. The nature of the cliff face and the platform both depend on the dip of the rocks. Predominantly horizontal sedimentary strata in Cobequid Bay, Cumberland Basin, Northumberland Strait and Cabot Strait result in near-vertical cliffs and smoothly sculptured intertidal platforms.

Where the structure is vertical, as in Bay of Fundy basalts or sedimentary strata west of Parrsboro, vertical cliffs may again be formed, but with a more irregular platform. Inclined strata may produce overhanging cliffs if the dip is landwards, as in northwest Cape Breton or in some Meguma-group slates on the Atlantic Coast. Features such as caves, arches and stacks are not common.

Where erosion is rapid, small variations in resistance produce such features but, by definition, they are short-lived. Conditions necessary for the formation of these erosional features are best developed in Cobequid Bay, where a dramatic arch is found at Salter Head and a series of stacks at Five Islands. Other stacks, "The Three Sisters," can be seen 10 km north of Cape Chignecto, at Eatonville. Caves eroded in the Halifax formation can be seen at "The Ovens" in Lunenburg County.

Many of the cliffs are unstable. In the glacial-till cliffs that border the Northumberland Strait, slumping, especially in the spring thaw, is the predominant destructive process. Where sedimentary rocks dip seawards, seepage may produce slides, as in Horton shales in northeast Cape Breton. The large-scale slides can also be seen in the schists north of Cap Rouge on the Cabot Trail in Cape Breton Highlands National Park. Large blocks detach themselves from the horizontally bedded cliffs of southeast Cape Breton and in the Bay of Fundy. Huge masses of rubble accumulate at cliff foots in Triassic basalt as frost shatters the vertical structure. The highest actively eroding cliffs in Nova Scotia, in Advocate Bay at Cape Chignecto, are 210 m in height. Other high cliffs are on Cape Split, 80 m, and Cape Blomidon, 180 m.

Intertidal Wave-Cut Platforms

As cliffs recede, a gently upward-sloping bench is cut into the bedrock. Waves cut this bench by removing cliff-foot debris, scouring fragments against the cliff foot and across the platform, through wave "quarrying" in storms, and by chemical disintegration in the spray zone. Since wave attack is controlled by the limits of tidal action, it is clear that the landward and seaward limits of the platform must also be determined by tidal heights. Wide platforms are found on more-exposed coasts with high-energy waves and on coasts with a large tidal range. However, in Nova Scotia the highest-energy coast, the

Atlantic Coast, is dominated by highly resistant rocks. For much of the Atlantic Coast, there is no marine platform and no cliffs. In contrast, the entire Bay of Fundy coast exhibits a wide wave-cut platform cut into both basalts and the less-resistant sedimentary rocks. East of Digby Gut, a wide, continuous platform has been cut into the basalts that form near-vertical cliffs at the landward edge of the platform, up to 30 m in height. In Cobequid Bay, the platform is even wider, reaching a maximum 6000 m at Lower Economy. The narrowest section in the Bay is 900 m at Burntcoat Head, presumably because the sandstone is more resistant in this area.

The great width and extreme rapidity with which this platform has been created, and is still being extended, are due to the large tidal amplitude. In most areas, the platform is partially obscured by sediments, which typically show the following sequence: gravel beach, mud veneer over gravel lag, exposed gravel lag, sand bars. These sediments are predominantly thin, less than 0.5 m. Gravel at the cliff foot forms a steep beach, which in places is being driven inland over salt marshes by the rising sea level. The mud veneer, in Cobequid Bay, is thin on the rock platforms and appears to thicken during the summer and to be thinned by winter ice. Near the low-water level, the mud becomes coarser and more extensive. Sometimes deep sand deposits obscure the platform. A gravel-lag deposit, from which the finer sediments have been washed out, underlies this entire sequence in some sections.

Wide intertidal platforms also dominate the Northumber land Strait coast from Merigomish to Baie Verte. Little research has been conducted on the formation of these platforms. Their presence is attributed to the unresistant bedrock. Sand bars are frequently found on these platforms, which have an almost imperceptible gradient.

Beaches

Beaches are wave-dominated deposits composed of a mixture of sand, gravel and other sizes of sediments. Sand beaches occur in sediment "sinks," in which all available material from a stretch of coast known as a "cell" is concentrated. Sand beaches often contain a significant gravel element at the higher levels. Very large particles in the boulder/cobble range are generally found close to their source, usually an eroding glacial-till cliff.

Concentrations of sand and gravel are formed under a great variety of circumstances and are the basis for the following beach classifications:

- Barrier beaches and baymouth bars
- Spits
- Tombolos
- Pocket beaches

Barrier Beaches and Baymouth Bars

These beaches are found only in areas with abundant unconsolidated sediments. Rising sea level erodes the beach and adjacent headlands, forcing a landward retreat of the beach. The adjacent headlands anchor the extremities of barrier beaches. Barrier beaches are well developed in eastern Cape Breton. The sandy beaches of Bridgeport Basin and Big Glace Bay Lake are sand-and-gravel barriers. The barrier beaches at Belfry and Fourchu on the Atlantic Coast are formed of pebbles, cobbles and sand. Large barrier beaches are found at Clam Bay, Martinique and Lawrencetown, at St. Catherines River Bay and elsewhere on the Atlantic Coast.

These are all composed largely of relict sands accumulated during deglaciation, rather than sands which are actively accumulating at present. Barrier beaches are largely absent from the Bay of Fundy and are found only in pronounced sediment traps in Northumberland Strait at, for example, Oak Island, Caribou Island and Merigomish Island.

On the more exposed Gulf of St. Lawrence coast, barriers are again well developed, especially at Antigonish and Pomquet and at Inverness and Pleasant Bay. As a result of their continuing landward migration, all these beaches undergo a constant cycle of erosion, failure and rebuilding. All stages of this process may be observed at various places in the province.

The Cadden Bay barrier beach has failed twice in storms during the last fifty years, a fact confirmed by aerial photography and indicated by the flat profile, absence of dunes and restricted vegetation on the beach at present. Inverness Beach by contrast, is well developed, with high, stable dunes and no evidence of overwash. Martinique is about to fail, with recent establishment of washovers, a high erosion face in the dunes, and invasion of dune vegetation by blowing sand, resulting in a reduction in diversity of shrubs and grasses in favour of dune grass.

The low wave energy of Northumberland Strait has not allowed the incipient barrier beaches to build above high-tide level, but has maintained an almost horizontal subtidal profile over large areas, as at Oak Island. At the eastern end of the Strait, which is more exposed, higher wave energy has allowed the emergence of the barrier beaches. The best example, at Merigomish, is also failing, with washover occurring in most major storms.

Spits

Spits are created by the movement of sediments along the shore by waves and, in the Bay of Fundy and Bras d'Or Lake, tidal currents. At points where there is an abrupt change of shoreline orientation or the currents diminish, sediment is deposited. Many of the barrier beaches have been initially formed by the coalescing of spits forming from opposite directions, or by the complete closure of a bay by material moving in one direction. Coalescing spits have formed the beaches at Advocate Harbour and West Apple River, Cole Harbour and South Harbour, Ingonish while materials moving from one direction have formed barriers such as Merigomish Island, Mabou Harbour and Martinique Beach.

Beaches forming on spits, including the recurved downdrift extremity, are uncommon. One of the best is at St. Anns Bay and another is in Barrington Bay. St. Anns is a good example of a spit that grew as a series of recurved beach ridges that are now drowned by the present beach. Probably the best classic examples of spits are on the north shore of Minas Channel and the supratidal elements of the barrier-bar system at Oak Island on the Northumberland Strait. The two spits extending westward from the island fail to form a permanent connection with another southward-extending spit on the mainland. Where currents meet on straight shorelines, spits may coalesce to form a triangular beach enclosing a small lagoon. These "barachois" ponds are especially common in Bras d'Or Lake where the incoming and outgoing tidal currents interact.

Tombolos

Tombolos are beaches formed in the lee of islands, where wave action is reduced. Frequently tombolos extend from the mainland shore to the island, forming a permanent connection. Many of the larger and better-known beaches are of this type: Partridge Island at Parrsboro, the former beach at Port Hood and the double tombolo at Point Michaud.

Pocket Beaches

On coasts with cliffs and where sediment is scarce, limited amounts of sand, gravel or cobbles may accumulate in indentations in the coastline. These beaches are always small and generally become smaller as the sea level rises. The fine sand beaches on Pennant Head in Halifax County are of this type, as well as many of the steep cobble beaches on the Atlantic Coast; for example, on Flying Point, Martinique. Beaches of this type are also common in northern Cape Breton; for example, Fishing Cove and Pigeon Point Cove in the national park.

General

While most of the beaches are eroding, retreating landwards and losing sand, there are some examples of accreting beaches. Where accretion is most rapid, a series of beach ridges may be built parallel to shoreline.

The best example is at Pomquet Beach, which appears to be still growing despite rising sea level. Conrod Island is also composed entirely of recently constructed beach ridges but is now eroding. Just to the west, Cole Harbour beach is also accumulating both sand and gravel at a rapid rate.

Changes in exposure to wave attack by erosion of offshore drumlins may result in changes to the orientation of beaches. This occurs in two main areas: Mahone Bay and Lobster Bay. One of the best-known examples is Yarmouth Bar, described by Goldthwait, "...here ...are three groups of concentric beaches, intersecting one another in such a way as to show they were built by shore drift coming first from one side and then from the other...". The beach is currently being pushed landwards over the salt marshes earlier developed in its protection. Artificial changes to the shoreline are usually the most likely reason for modern beach construction.

Dunes

Nowhere in mainland Nova Scotia are sand dunes exceptionally well developed, compared with areas surrounding the southern Gulf of St. Lawrence in P.E.I. and Newfoundland. This may be attributed to a scarcity of sand on the higher-energy coasts and to lower-energy environments in areas where sediment is abundant.

Probably the largest mainland dunes are those at South-West Port Mouton, reportedly 15 m high. Elsewhere on the high-energy Atlantic Coast, dunes are found at Rissers Beach and in Barrington Bay. On many other beaches, low ridges colonized by Marram Grass are commonly referred to as dunes, but many of these backshore ridges have been thrown up by the waves with only minor additions of windblown sand.

Only a few, very low dunes are found in the Bay of Fundy, and in the Northumberland Strait sand dunes are restricted to the more exposed easterly sections. Features referred to as dunes at Caribou Island, Merigomish Harbour, Antigonish and Pomquet are more properly classified as wave-created beach ridges. True sand dunes are found at Mabou Harbour, Port Hood, Aspy Bay and Inverness, where copious sand supply coincides with strong on-shore winds. Large sand dunes are associated with the beaches near Mavillette, such as at Bartletts Beach.

Islands

Islands, areas of land surrounded by water, are characteristic of submerging coastlines. There are many hundreds of islands around the coast of Nova Scotia, particularly along the Atlantic Coast. They range in size from large to small reefs exposed only at low tide. Islands may be composed of bedrock, glacial till or sand. Bedrock islands are most common, particularly in areas of resistant rock, such as the granite and quartzite of the Atlantic Coast but also of softer sedimentary rocks and basalt in the Northumberland Strait and Bay of Fundy.

Glacial-till islands occur where drumlin fields have been drowned by rising sea level. This particularly occurs in Mahone Bay and St. Margarets Bay and east of Halifax Harbour. Drumlins erode more rapidly than bedrock, gradually being reduced to boulder reefs. The eroding till provides ample sediment supply for the development of sandy beaches and cobble-shore habitats.

Islands of sand are generally low lying and change their shape as a result of water currents and storms. Larger islands, such as Pomquet Beach and Sable Island are more stable and have developed sand-dune and coastal-forest vegetation. Islands have a high value as wildlife habitat, particularly for breeding seabirds and waterfowl but also for breeding seals. The value is enhanced by the degree of isolation and the state of residual vegetation. Vegetation is characteristically sparse on sand-dune habitats, but on large islands includes White Spruce coastal forest. Bedrock islands are most suitable for Eider Ducks, terns, gulls, and also Blue Heron, osprey and cormorants if there are sufficient trees and the site is not too far offshore. Some of Nova

Scotia's bedrock islands are also suitable for nesting Black Guillemot. Drumlin islands and bedrock islands with sufficient soil provide habitat for Puffins and petrels, which build their nests in burrows. Sand islands are important for shorebirds such as the endangered Piping Plover. There are relatively few islands with cliffs suitable for alcid nesting, the most important being Bird Islands in Units 531 and 915 and St. Paul Island.

Another important biological feature of islands is the isolation and adaptation of species of plants and animals that have been separated from other populations by rising sea level. Classic studies of "island ecology" have provided important insights into the mechanisms of adaptation and evolution. Species of interest on islands of the coast of Nova Scotia include Geum peckii, land snails Cepaea hortensis, Redback Salamanders, Garter Snakes and small mammals. The importance of islands to wildlife is reflected in the establishment of several protected areas of islands, *e.g.*, Eastern Shore Islands Wildlife Management Area and single islands, *e.g.*, Pearl Island Wildlife Management Area. Islands that are significant as wildlife habitat are shown in the Atlas of Important Freshwater Wetlands and Coastal Wildlife Habitats for Nova Scotia. Other land uses have included sites for lighthouses, sheep grazing, farming/fishing stations or communities and, more recently, vacation cottage or resort development.

Sable Island

In sharp contrast to the mainland, Sable Island has an abundance of sand and strong winds. As a result, dunes over 25 m in height are found and continue to be formed. Most dunes are 5–15 m in height and occur in two ridges oriented east–west, parallel to the shoreline.

These two ridges are gradually moving together, obliterating the once-continuous pond between them. The northern ridge is higher and probably older than the southern ridge, which is less than 10 m high. Materials are entirely sand; no native pebbles have ever been found. It appears that the central part of the island is essentially stable, but the spits at either end are very mobile and move back and forth with storm activity.

Controversy surrounds the question of the ultimate survival of the island. Some think that storms will reduce the island to a submarine bar. However, recent investigation indicates that while the island has been becoming narrower through loss of sand from the south side, it has been increasing in height relative to rising sea level. Perpetual movement of the sand dunes by the persistent strong winds has prevented the establishment of shrubs. Roland records that the 80,000 shrubs planted in 1901 had dwindled to 77 in 1913 and one in 1952. The contribution of the Sable Island horses to this loss of shrubs and increased mobility of sand is also a matter of some controversy.

Tidal Deltas

In inlets to lagoons, and in estuaries characterized by strong tidal currents, reworked marine sediments are commonly found on both the seaward and landward sides of the inlet where tidal currents abruptly lose velocity.

Inlets in barrier beaches, where sediments are plentiful, usually exhibit tidal deltas, particularly on the landward side. Cole Harbour has almost been filled with sands in a very large flood-tidal delta. On the seaward side, the ebb-tidal delta is much smaller, as the sediments are continually removed by wave action. Changes to the channel immediately result in current changes, which are soon reflected in the form of the tidal delta.

Tidal deltas rarely emerge above the highest high tide, as their deposition is dependent on tidal currents. While tidal deltas are a normal feature of breached barrier beaches, they are best developed in Cobequid Bay where exceptionally strong tidal currents occur.

Flood-tidal currents are strongest along the shorelines, while the ebb tide dominates the central channel in the Bay. The entire sand body reaches 5 km in width and is up to 25 m thick. Particle size decreases towards the head of the Bay, but fine sand remains the dominant sediment well into the Salmon River and Shubenacadie River estuaries. Sand is abruptly replaced by mud in embayments and in the estuaries. Six main sand bars constitute the tidal delta in Cobequid Bay and five in the Avon River estuary. The sand bars are elongated, up to 15 km in length, approximately aligned with the tidal currents. Sand waves and megaripples dominate the surfaces of these bars.

River Deltas

The very limited amount of sediment carried by rivers and a large tidal range preclude the development of deltas in the estuaries of most Nova Scotia rivers. In Bras d'Or Lake, the small tidal range has permitted the development of small deltas at the mouths of the larger rivers, particularly in Nyanza Bay and Whycocomagh Bay. These river deltas show the typical features: levees lining the main channel, which extends beyond the distributory channels; little emergence above the lake level; and emergent vegetation over the entire delta.

Mudflats

Mud flats are composed of predominantly clay-sized particles and water, and are distinguished from salt marshes by a lack of vegetation. A more accurate sediment description in the Bay of Fundy, where the largest mud flats occur, would be muddy, sandy silt-flats. The lower mud flats contain more sand and silt and less water than the upper mud flats. Newly deposited flats, such as that north of the Windsor causeway are more fluid than more stable flats, which contain only about 20 per cent water, and have virtually no shear strength. In some circumstances the mud flats are a veneer over intertidal rock platforms, but in sheltered bays and inlets, deep deposits are found, up to 25 m deep in places.

Vegetation is able to colonize only the mud flats near the higher tide levels. When this happens, further fine sediments are trapped by the vegetation, building the marsh up to within a metre of the higher high-tide level. The mud flat proper forms below, about 4 or 5 m above the mean tide level in the Bay of Fundy. Large mud flats are found in the southern Minas Basin both east and west of the Avon River estuary and in Cumberland Basin. It has been shown that the activities, especially secretions, of marine organisms, help to stabilize the mud flats. Such activity may influence the mud flat to a depth of 30 cm.

Fine sediments composing the mud flat are derived from eroding cliffs and carried in to the upper Bay by flood-tidal currents. The ebb-tide current is unable to carry out the same amount of sediment. Concentrations of sediment in the water have been measured at up to 10 g/litre in Cumberland Basin and up to 40 g/litre in tidal rivers.

Salt Marshes

Bay of Fundy salt marshes can be distinguished from salt marshes on other coasts by their height, sediment distribution and organic matter. The Bay of Fundy salt marshes are mud flats raised and stabilized by the effects of colonizing vegetation. The marsh surface is almost horizontal, slightly more than a metre below the highest high-tide levels.

As the tidal height is elevated in the estuaries, the salt marshes may be formed as high as 4.5 m above chart datum. Reclaimed marshes in the Annapolis, Salmon, Shubenacadie, Avon, Minas Basin and Cumberland Basin areas vary in height from 8.0 m to 14.6 m, depending on local tidal conditions. The relationship between tidal heights and marsh levels is shown in Table.

Table. Relationships of Tidal Heights and Marsh Levels (in metres) in Selected Bay of Fundy Marshes.

	Digby/Annapolis	Burncoat Head	Hantsport
Large Tide	9.33	15.79	15.29
Mean Tide	7.99	13.56	13.19
Mean Sea Level	4.54	7.53	7.07
Tidal Marsh	8.13	14.59	14.04

Organic matter is a small component of these marshes, which are composed of fine sands, silt and clay-sized particles. The colour of the marsh soils indicates the type of sediments. Bay of Fundy marshes can be classified as "red," with better drained, coarser sediments, and "blue," with finer, poorly drained sediments.

Formation of the marshes has been likened to a river in reverse. The tide flows in through the main channel and then spills over and branches into a series of smaller channels. Levees of coarser, sandy sediment are built on the channel margins, while the fine material is deposited further from the channels. On the larger marshes, fresh water may be trapped at the lower marsh elevations, adjacent to the upland on the landward side of the marsh. Rising sea level has accentuated this elevation difference, often forming a plug at the entrance to a river valley, damming fresh water inland and allowing the formation of peat deposits. Such situations are found in the Cornwallis River valley and on the Chignecto marshes. The large Bay of Fundy marshes are unusual in that the source of the sediments is chiefly marine; elsewhere, large salt marshes are usually formed from river-borne sediments. A copious supply of marine sediment derived from eroding drumlins in Lobster Bay, together with a large tidal range, has also caused the formation of extensive marshes in the Pubnico area. Marshes on the Atlantic Coast and Northumberland Strait are smaller, have accumulated more slowly, and usually contain much greater amounts of organic matter. Rising sea level is driving beaches inland over the marshes at numerous locations, and in places compressed peat deposits outcrop seaward of the beach. At the inland limit, the coastal spruce–fir forest is characteristically being inundated by the sea, and organic deposits are slowly spreading into the forest. A fringe of bleached coniferous tree trunks killed by salt water is a common backdrop to Nova Scotia salt marshes.

Karst

The term "karst" stems from the region Krs in Slovenia, now in northwestern Yugoslavia, which is typified by stony barren rock. The Indo-European word "kar" for rock and the Italian word "carso" evolved to the Germanized term"karst," which is now the accepted term for solution-derived landscapes like those just north and east of the Adriatic Sea.Karstification is the geologic process of differential chemical and mechanical erosion by water on soluble bodies of rock, such as limestone, dolomite, gypsum, or salt, at or near the Earth´s surface. Karstification is exhibited best on thick, fractured, and pure limestones in a humid environment in which the subsurface and surface are being modified simultaneously. The resulting karst morphology is usually characterized by dolines, hums, caves, and a complex subsurface drainage system.

Factors in Karst Development

The development of karst terrain depends on the interplay of at least seven important factors in varying degrees. These are: lithology, structure, relief, hydrology, climate, vegetation, and time.

Lithology

Several lithologies are susceptible to karstification, but limestones and dolomites, owing to their solubility and nature of resistance and widespread distribution, are overwhelmingly dominant. 75 per cent of the Earth's surface is covered with sedimentary rocks, and of that, 10 to 20 per cent consists of limestones or dolomite. Karst morphology can occur on carbonates with less purity than 80 per cent, but generally the purer the limestone, the better the development of karst morphology.

For example, the Cretaceous and Early Tertiary Dinaric karst limestones of Yugoslavia are 95 to 100 per cent pure, and white limestone in Jamaica seldom exhibits less than 98 per cent purity.

Numerous compositional variations can exist in carbonates, and many of these are discussed in Sweeting and Jennings. In addition to the composition of limestone, the thickness of individual beds, the nature of interbeds, especially shaly beds, and lateral facies variations affect the style and degree of karstification. In the Dinaric Mountains, the depth of surface karstification is sometimes limited by impermeable beds which have aided the development of broad flat-floored valleys called poljes.

Structure

Structure is used here in the structural geologic sense and is concerned with the attitude and deformational effects of bedrock. Limestones and dolomites at or near the surface tend to deform by brittle fracture.

This tendency to form complex joint sets is directly responsible for the secondary permeability required for the development of subsurface solution drainage and subsequent three-dimensional cave development. Vertical fractures usually manifest themselves at the surface and focus the solution processes along them. The influence of fracturing is paramount in southeast China, where thousands of square kilometers are affected by several consistent sets of joints. Folding, in addition, may have the effect of isolating watersheds and directing ground-water flow. This has occurred in the Dinaric karst of Yugoslavia.

Large-scale structures or tectonics not only have led to the development of specific landforms like poljes but have also influenced the rates and degree of karstification. Most poljes, for example, are associated with boundary faults.

Uplift and deformation of a carbonate plateau accelerates the dissolution of the limestone because:

- Fracture density increases,
- Any elevation rise is usually accompanied by an increase in precipitation, and
- Increased relief increases piezometric surface gradients.

These conditions can:

- Lead to solution-enlarged conduits along fracture planes,
- Impart to the limestone a high hydraulic conductivity, and
- Cause rapid fluctuations in the water table with accompanying accelerated solution.

Relief

Topographic relief is the elevation difference between the highest and lowest points on the surface for a given area. The Julian Alps in the inner region of Yugoslavia have a topographic relief near 2000 m.

This mountainous area is associated with an abundance of potholes and few caves. Both characteristics are thought to be a consequence of this high relief. Related to relief is the depth of the vadose zone, which is the vertical distance between the surface and the water table. In Florida, the vadose depth is small, as is the topographic relief. However, throughout low sea level stands during the Pleistocene, the increased depth of the vadose zone probably led to the development of a more extensive and efficient subsurface drainage system with an increase in dissolution along fractures and fissures. With the absence of the buoyancy effect of a high water table surface, the formation of collapse dolines may have been accelerated. Even today the effects of water table drawdown due to excessive pumping have rendered some urban areas susceptible to the sudden and unexpected collapse of buried solution cavities. Such a collapse occurred in Winter Park, Florida, in May 1981.

Hydrogeology

Hydrogeology is concerned with the interrelation ships of geologic materials and processes with water. The karstification process involves not only the solution of limestone but also the sometimes overlooked mechanical erosion of limestone. Because the solution chemistry of carbonates is well known only a brief summary will be presented here.The dissolution of limestone involves three principal components: carbon dioxide, water, and calcium carbonate.

Initially, atmospheric carbon dioxide diffuses into the moisture within the air or soil and simultaneously becomes hydrated to form carbonic acid:

$$CO_2 + H_2O \leftrightarrow H_2CO_3$$

In the presence of limestone, carbonic acid dissolves the calcite. The reaction is often presented as:

$$\underset{\text{Limestone}}{CaCO_3} + CO_2 + H_2O \leftrightarrow \underset{\text{Dissolved limestone}}{C_a^{++} + 2HCO_3^-}$$

The only water that is capable of dissolving limestone is aggressive water, which is water undersaturated with respect to dissolved carbonate. The time required for aggressive water in karst regions to be neutralized or to reach saturation equilibrium varies considerably, depending on a number of factors such as temperature, turbulence, variations in the partial pressure of carbon dioxide, dilution, presence of other acids, and surface area of limestone. To achieve equilibrium may require several days; however, laboratory studies suggest that most of the limestone dissolution resulting from an influx of fresh aggressive water may occur within minutes to a few hours. Additional acids, such as organic acids from soils and most recently sulfuric and nitric acids from acid rain, will contribute to the dissolution of carbonate rocks.

Once saturation equilibrium is achieved, any change to warmer water temperatures or decreases in the partial pressure of carbon dioxide can drive equation 7-2 to the left, resulting in the deposition of calcium carbonate. Erosion rates have been calculated for many karst regions around the world and vary widely even within the same climatic zone. Although these erosion rates are useful for comparative purposes, they are not indicative of the duration intensity of the erosion process. The dissolution of the limestone does not take place at a constant rate, but fluctuates with storm events that flood the karst surface and subsurface with undersaturated water. In other words, the actual dissolution occurs episodically with maximum peaks corresponding to surges in the karst water flow. In Jamaica, intense storm events result in subsurface flooding, and because the existing conduit system is not of sufficient size to handle excessive infiltration rates, backfilling of caves and solution channels occurs with fresh undersaturated water. These flood events can be of the single-storm variety or may be seasonal, as with the winter flooding of poljes in Yugoslavia.

Table. Erosion Rates for Image Areas Discussed

Location	**Erosion Rate ($m^3/km^3/yr$)**	**Climate**
Florida USA	35,5,8	Tropical
Jamaica	69,39,86,96	Do
Yucatan, Mexico	30,16	Do
Guangxi, S.China+	120-300 (surface)	Do
Dinaric karst, Yugoslavia	43,63,55,67	Temperate
S. Nullarbor Plain, Australia±	5 (surface)	Semiarid

Because of the unusually high permeabilities and the periodic flooding associated with most karst regions, the water table correspondingly exhibits drastic fluctuations. This has been observed by many spelunkers. The water table in karst rock is frequently difficult to define, especially in free-flow carbonate aquifers that display a complex multitude of enlarged open conduits. The water table can occur at considerable depths below the land surface, especially in mountainous regions like the Dinaric Mountains, Yugoslavia.

Water can also be perched under solution depressions above the water table or it may be discontinuous if the permeability is principally fracture-controlled. Zones with few or no fractures yield little or no water, in contrast to dense fracture zones that can potentially supply abundant water.

Climate

The best-developed karst regions of the world are found in tropical and temperate environments. Considerable discussion has evolved as to whether tropical or temperate environments are more conducive to karstification. Lehmann, a strong advocate of climatic geomorphology, asserts that solutional erosion rates are more intense in the tropics because of greater rainfall and carbon dioxide production in soils at higher temperatures. In contrast, Corbel contends that the intensity is greater in cold humid climates where larger quantities of carbon dioxide can be absorbed by water. The higher the carbon dioxide absorption in water, the stronger is the resultant carbonic acid. However, because Corbel overlooked the rates of reaction, which slow down at lower temperatures, the levels of CO_2content may be irrelevant.

The absorption of carbon dioxide is temperature-dependent so that cooler water for a given volume can absorb larger quantities of carbon dioxide. If the water temperature rises from 0 to 35°C, the carbon dioxide saturation level will decrease between one-third to two-thirds. Therefore, waters in cold climates can hold greater quantities of carbon dioxide in the ionic state, thus being more effective in dissolving limestone. However, more important than temperature is the partial pressure of carbon dioxide, which at sea level is relatively constant at 0.035 per cent. The partial pressure of carbon dioxide in soil air can exceed that of the atmosphere from ten to several hundred times. As a consequence, the presence and nature of soil cover can be far more important in karstification intensity than atmospheric air.

A study by Smith and Atkinson of comparative karst data from around the world indicates that mean annual run-off, rather than precipitation, is the principal parameter directly related to erosion rate. Temperature is believed to be important only with regard to its influence on supporting a continuous cover of soil and vegetation.

Vegetation

Arid karst regions, such as the Nullarbor Plain in Australia, are associated with sparse vegetation, thin soils, and therefore, a slowed rate of surface karstification. In contrast, if the existing soil and climate can support a dense vegetation cover, an intensified surface and subsurface karstification occurs, as observed in Jamaica and south China. Biogenic carbon dioxide, introduced to the soil by plant root systems and bacterial decay, is regarded as the most important control of solution erosion of limestones. Generally, where vegetation is most lush, the highest concentration of moisture, the highest biogenic activity, and the highest production rates of carbon dioxide occur. Under these conditions and free water circulation, the highest solution erosion rates can be achieved. Such conditions are found in both Jamaica and south China. In Jamaica, the forest canopy has been cited as one of the prime factors responsible for the overdeepening of cockpits.

Time

The early karst studies of Cvijic and Grund at the turn of the 20th century were strongly influenced by the cyclic evolutionary teachings of Davis and Penck. Thus, karst topography has been thought to evolve through stages of development beginning with fluvial action and the initial formation of dolines. As the surface drainage is slowly captured by swallow holes and the developing subsurface drainage, dolines coalesce to form uvalas, which, in turn, expand to poljes. This sequence is now considered to be antiquated, and just as to Roglic, poljes are structurally controlled and are not related to uvalas.

It is questionable whether any two karst regions anywhere in the world have evolved sequentially in the same manner or even display identical morphological forms. Each karst region evolves within its unique combination of dynamic and static factors. However, this does not preclude the comparison of areas of similar morphology in an attempt to evaluate common factors. From the examples of Landsat imagery in this stage, it becomes apparent that each region has developed its own morphologic signature. To understand the evolution of a given karst terrain, it is necessary to comprehend the roles of length of time, style, and intensity of the energy flow through the hydrologic system. Those karst regions with the largest potential energy, related to the thickness of the vadose zone, and kinetic energy, or flooding, will also exhibit a higher energy morphology more rapidly and more fully than regions of lower total energy. High-energy morphology can be manifested in the form of larger caves, increased relief, density, and magnitude of dolines, hums, flood deposits, and a more extensive underground solution network.

Karst from Space

Within the realm of geomorphology, each geologic process leaves its own imprint on the Earth's landscape, and each process will develop its own characteristic assemblage of landforms. For karst landscapes, this imprint is expressed as solution morphology on the regional scale. Because of the relatively large ground resolution distance of 79 m, many karst landforms cannot be discerned from Landsat MSS images; only the larger solution and fracture-controlled dolines and uvalas can be recognized.

Thus, the advantage of the space perspective is not the identification or the recognition of individual landforms, but the collective pattern and texture they impart to a region of hundreds or thousands of square kilometers. This can occur only where given geologic processes and materials, such as carbonates, dominate a large region for a sufficient length of time. The Landsat imagery discussed in this stage has been chosen from the known karst regions of the world. Some well-researched regions are not included either because good imagery has not been available or because man's overprint is so dominant as to obscure natural topographic patterns.

A cursory examination of all the karst images reveals a number of features specifically associated with the solution process:

- Lack of drainage patterns
- Dimpled pock-marked texture
- Fracture sets with developed relief
- Uniformly scattered residual karst hills
- Circular basins or lakes
- Large flat karst plains and poljes

Most important is the absence of a well-developed integrated surface drainage. Indeed, for the Nullarbor Plain, Australia, and Yucatan, Mexico, it is difficult to identify any recognizable drainage. Major rivers can be seen to incise the south China carbonate plateaus, but there is a distinct lack of well-developed tributaries draining into the river channels.

Most karst areas exhibit a relatively homogeneous pock-marked or dimpled texture. In south China, this pattern is overprinted with solution-incised joint sets. In Florida, the pattern is expressed as numerous large water-filled collapsed basins. Joint sets are not recognized in all karst areas at Landsat scale.

This does not preclude the presence of joints, but merely points out that they may be obscured because:

- Joints are too closely spaced,
- Solution relief of joints is slight,
- Dense vegetation cover is present, or
- Joints may indeed be absent or sparse.

What accounts for the large differences observed in karst regions is not variation in the karst process, but variation in lithology, structure, and the overprint of other processes like fluvial, tectonic, eolian, and glacial, and variations in the energy flow through the karst system through time.

COASTLINES

Coastlines are boundaries between land and water that surround Earth's continents and islands. Scientists define the coast, or coastal zone, as a broad swath (belt) of land and sea where fresh water mixes with salt water. Land and sea processes work together to shape features along coastlines. Freshwater lakes do not technically have coastal zones, but many of the processes and features found along ocean coastlines also exist in large lakes.

Coastal zone features

All coastlines include a thin strip of land that is submerged at high tide and exposed at low tide, called the shoreline. The coastal zone, however, extends far inland from the shore, across lowlands called coastal plains, and far seaward to the water depth where ocean waves do not reach the seafloor. The coastal zone includes lagoons, beaches, estuaries, tidal wetlands, tidal inlets, river deltas, barrier bars and islands, sand bars, and other shallow–water ocean features.

- *Lagoons*: Shallow, salt–water bays between barrier islands and the mainland.
- *Beaches*: Sand deposits along shorelines. Intense waves wash fine-grained mud from coastal sediments (particles of sand, gravel, and silt) leaving only sand-sized grains of resistant minerals like quartz and calcium carbonate. Beaches are common on the seaward side of barrier islands where wave energy is intense.
- *Estuaries*: The mouths of rivers and streams that receive a pulse of saltwater with the tides.
- *Tidal wetlands (flats)*: The broad areas of marshy wetlands around lagoons and estuaries that flood with salt water during high tides.
- *Tidal inlets*: Openings through which water and sediment are washed in and out of lagoons by daily tides.
- *Deltas*: Deposits of sediments at the mouths (ends) of rivers that flow into the ocean.
- *Barrier bars and islands*: Long mounds, or bars, parallel to the shore into which near-shore ocean currents carry and deposit sand. Eventually, some barrier bars grow tall enough to stay exposed at high tide and become barrier islands. The outer banks of North Carolina as well as Galveston, Mustang Island, and South Padre Island in Texas are examples of barrier islands.
- *Sand bar*: A ridge of sand in rivers or along the coast built up by water currents.

Processes that shape coastlines

The coastal zone is constantly changing. Salt water rushes through tidal inlets into bays and estuaries twice daily. Waves, currents (steady flows of water in a prevailing direction), tides, and storms reshape coastal features over days, weeks, and months. Coastlines move landward and seaward as global sealevel rises and falls over hundreds and thousands of years. All coastlines are at least somewhat affected by waves and tides.

Waves straighten uneven shorelines by eroding (wearing away) points that extend into the ocean and depositing sediment in bays. They also generate strong, shallow currents that carry and deposit sediment parallel to the shore. Long shoreparallel features like barrier islands, spits (small strips of land that jut out into the sea), and sand bars border coasts where waves are the dominant force. Tides move sediment and water in and out across the shoreline, and tide–dominated coasts have features like tidal inlets, natural jetties (protective rock barriers), and funnel-shaped estuaries that form a 90° right angle to the shore. Most coastlines are shaped by both waves and tides, and have some parallel and perpendicular features.

Types of coastlines

All coastlines are affected by waves, tides, storms, and currents, and every coast includes a shoreline. There are, however, many different types of coastlines. Some coastlines receive large amounts of sand and mud from rivers. Others accumulate the skeletal remains of animals like corals and shellfish. In some places, waves are eroding coastlines that are rising from the sea.

Depositional Coastlines

Coasts that receive a steady supply of sediment are called depositional coastlines. Rivers like the Mississippi in the United States and the Nile in Egypt erode sediment from continental interiors and deposit it in huge deltas at their mouths. Waves, currents, and tides spread sediment into thin layers on the submerged continental shelf (the shallow seabed that stretches from the shore to the deeper ocean water). Over time, the weight of the sediment presses down on the edge of the continent, creating space for more sediment. New layers build on to the edge of the continent, and the coast moves seaward. Depositional coastlines typically encompass a broad coastal plain, and a complex shoreline that includes long, wide beaches. These coastlines are almost flat, causing salt water to move far inland across coastal plains and up rivers during high tide.

The Mid–Atlantic and Gulf of Mexico coasts of the United States are depositional coastlines. Waves, tides, and currents sort and distribute incoming sediment into distinctive features along depositional coastlines. Some of these features include: barrier bars and islands, tidal inlets, lagoons, beaches, estuaries, and tidal wetlands. Depositional coastlines also develop where plants and animals called carbonates live in clear, sunlit water away from river deltas. Carbonates like corals and shellfish have skeletons and shells made of the hard mineral calcium carbonate.

The sediment supply on carbonate coastlines comes from the skeletal remains of the animals and plants that live there. Corals build giant ridges of rocks called reefs up from the seafloor. Florida and the Bahamas have carbonate coastlines.

Erosional Coastlines

Erosional coastlines occur where huge sections of the Earth's crust called tectonic plates lift out of the sea. These coastlines are common along far northern coastlines that are bouncing back after being weighed down by thick ice sheets, and along coasts where tectonic plates meet. Erosional coastlines are the norm in Maine and eastern Canada, along the west coast of North America, and in Scandinavia. Along these coasts, waves pound rocky shorelines and cut into the bottoms of cliffs. The shore retreats as blocks of rock and sediment fall into the sea. Isolated remnants of sea cliffs, called stacks, are left standing in the sea. All of these features are caused by movements of plates and wave action over time. The process of shoreline retreat claims much expensive real estate along erosional coastlines.

Life in the coastal zone

The plants and animals that live in the coastal zone have adapted to its cycles of change. Residents of tidal wetlands tolerate twice–daily drowning and drying. Fish in estuaries and lagoons adjust to large changes in water salinity (saltiness). Plants grow with their roots in salt water and can survive burial by shifting beach sand and river mud.

Many ocean and land animals spend their early lives in coastal wetlands where there is shelter and plentiful food before moving to dry land or the open ocean as adults. Coastlines are also home to about twothirds of Earth's human population. People continue to work to understand the nature of coastal zones in order to protect coastal populations from their hazards (storms, waves, floods, erosion), but also to protect coastlines from the damaging effects of everyday human activities, such as eroding sand dunes by climbing them, or generating pollution.

CAUSES OF COASTAL HAZARDS

The population that lives along or near our coastlines are an extremely vulnerable population. There are numerous issues facing our coastlines and there are two main categories that these hazards can be placed under, National disasters and Human disasters. Both of these issues cause great damage to our coastlines and discussion is still ongoing regarding what standards or responses need to be met to help both the individuals who want to continue living along the coastline, while keeping them safe and not eroding more coastline away.

Natural disasters are disasters that are out of human control and are usually caused by the weather.

Disasters that include but are not limited to; storms, tsunamis, typhoons, flooding, tides, waterspouts, nor'easters, and storm surge. Human disasters occur when humans are the main culprit behind why the disaster happened. Some human disasters are but are not limited to; pollution, trawling, and human development. Natural and human disasters continue to harm the coastlines severely and they need to be researched in order to prepare/stop the hazards if possible.

The populations that live near or along the coast experience many hazards and it affects millions of people. Around ten million people globally feel the effects of coastal problems yearly and most are due to certain natural hazards like coastal flooding with storm surges and typhoons. A major problem related to coastal regions deals with how the entire global environment is changing and in response, the coastal regions are easily effected.

Storms, Flooding, Erosion

Storms are one of the major hazards that are associated to coastal regions. Storms, flooding, and erosion are closely associated and can happen simultaneously. Tropical storms or Hurricanes especially can devastate coastal regions. For example, Florida during Hurricane Andrew occurred in 1992 that caused extreme damage.

It was a category five hurricane that caused $26.5 billion in damages and even 23 individuals lost their lives from the storm. Hurricane Katrina also caused havoc along the coast to show the extreme force a hurricane can do in a certain region. In almost all cases, storms are the major culprit that causes flooding and erosion. Flash flooding is caused by storms that occurs when a massive amount of rainfall comes down into an area over a short period of time. Where as a storm surge, which is closely related to tropical storms, is when the wind collects and pushes water towards low pressure or inland and can rise rapidly. It is an offshore rise of water and overall creates a higher sea level that rises and is pushed inland. The amount of rise or fall of storm surge depends greatly on the amount and duration of wind and water in a specific location. Also if it occurs during a high tide it can have an even greater effect on the coast.

Almost all storms with high wind and water cause erosion along the coast. Erosion occurs when but not limited to; along shore currents, tides, sea level rise and fall, and high winds. Larger amounts of erosion cause the coastline to erode away at a faster rate and can leave people homeless and leave less land to develop or keep for environmental reasons. Coastal erosion has been increasing over the past few years and it is still on the rise which makes it a major coastline hazard. In the United States, 45 percent of its coast line is along the Atlantic or Gulf coast and the erosion rate per year along the Gulf coast is at six feet a year. The average rate of erosion along the Atlantic is around two to three feet a year. Even with these findings, erosion rates in specific locations vary because of various environmental factors such as major storms that can cause major erosion upwards to 100 feet or more in only one day.

Pollution, Trawling, Human Development

Pollution, trawling, and human development are major human disasters that effect coastal regions. There are two main categories related to pollution, point source pollution, and nonpoint source pollution. Point source pollution is when there is an exact location such as a pipeline or a body of water that leads into the rivers and oceans. Known dumping into the ocean is also another point source of pollution. Nonpoint source pollution would pertain more to fertilizer runoff, and industrial waste. Examples of pollution that effect the coastal regions are but are not limited to; fertilizer runoff, oil spills, and dumping of hazardous materials into the oceans. More human acts that hurt the coastline are as follows; waste discharge, fishing, dredging, mining, and drilling. Oil spills are one of the most hazardous dangers towards coastal communities. They are hard to contain, difficult to clean up, and devastate everything. The fish, animals such as birds, the water, and especially the coastline near the spill. The most recent oil spill that had everybody concerned with oil spill was the BP oil spill.

Trawling hurts the normal ecosystems in the water around the coastline. It depletes all ecosystems on the ocean floor such as, flounder, shellfish, marsh etc.. It is simply a giant net that is drug across the ocean floor and destroys and catches anything in its path. Human development is one of the major problems when facing coastal hazards. The overall construction of buildings and houses on the coast line takes away the natural occurrences to handle the fluctuation in water and sea level rise. Building houses in pre-flood areas or high risk areas that are extremely vulnerable to flooding are major concerns towards human development in coastal regions. Having houses and buildings in areas that are known to have powerful storms that will create people to be in risk by living there. Also pertaining to barrier islands, where land is at risk for erosion but they still continue to build there anyway. More and more houses today are being taken by the ocean, look at picture above.

INTEGRATED COASTAL ZONE MANAGEMENT

Integrated coastal zone management means the integration of all aspects of the coastal zone; this includes environmentally, socially, culturally politically and economically to meet a sustainable balance all around. Sustainability is the goal to allow development yet protect the environment in which we develop. Coastal zones are fragile and do not do well with change so it is important to acquire sustainable development. The integration from all views will entitle a holistic view for the best implementation and management of that country, region and local scales.

The five types of integration include integration among sectors, integration between land and water elements of the coastal zone, integration amount levels of government, integration between nations and integration among disciplines are all essential to meet the needs for implementation. Management practices include

1. maintaining the functional integrity of the coastal resource systems,without disrupting the environment
2. reducing resource-use conflicts, by making sure resources are used adequately and sustainably,
3. maintaining the health of the environment, which means to protect the ecosystems and natural cycle,
4. facilitating the progress of multisectoral development, which means allowing developers to develop within standards.

These four management practices should be based on a bottom-up approach, meaning the approach starts from a local level which is more intimate to the specific environment of that area. After assessment from the local level, the state and federal input can be implemented. The bottom-up approach is key for protecting the local environments because there is a diversity of environments that have specific needs all over the world.

Adaptive management

Adaptive management is another practice of development adaptation with the environment. Resources are the major factor when managing adaptively to a certain environment to accommodate all the needs of development and ecosystems. Strategies used must be flexible by either passive or active adaptive management include these key features:

- Alterative decision-making (evaluating results and adjusting actions on the basis of what has been learned)
- Feedback between monitoring and decisions (learning process)
- Explicit characterization of system uncertainty through multi-model inference (experimentation)
- Embracing risk and uncertainty as a way of building understanding (trial and error)

To achieve adaptive management is testing the assumptions to achieve a desired outcome, such as trial and error, find the best known strategy then monitoring it to adapt to the environment, and learning the outcomes of success and failures of a project.

Mitigation

The purpose of mitigation is not only to minimize the loss of property damage, but minimize environmental damages due to development. To avoid impacts by not taking or limiting actions, to reduce or rectify impacts by rehabilitation or restoring the affected environments or instituting long-term maintenance operations and compensating for impacts by replacing or providing substitute environments for resources Structural mitigation is the current solution to eroding beaches and movement of sand is the use of engineered structures along the coast have been short lived and are only an illusion of safety to the public that result in long term damage of the coastline. Structural management deals with the use of the following: groins which are man-made solution to longshore current movements up and down the coast.

The use of groins are efficient to some extent yet cause erosion and sand build up father down the beaches. Bulkheads are man-made structures that help protect the homes built along the coast and other bodies of water that actually induce erosion in the long run.

Jetties are structures built to protect sand movement into the inlets where boats for fishing and recreation move through. The use of nonstructural mitigation is the practice of using organic and soft structures for solutions to protect against coastal hazards. These include: artificial dunes, which are used to create dunes that have been either developed on or eroded. There needs to be at least two lines of dunes before any development can occur. Beach Nourishment is a major source of nonstructural mitigation to ensure that beaches are present for the communities and for the protection of the coastline. Vegetation is a key factor when protecting from erosion, specifically for to help stabilize dune erosion.

COASTAL ENVIRONMENTS

There are many different types of environments along the coasts of the United States with very diverse features that affect, influence, and mold the near-shore processes that are involved. Understanding these ecosystems and environments can further advance the mitigating techniques and policy-making efforts against natural and man-made coastal hazards in these vulnerable areas.

The five most common types of coastal zones range from the northern ice-pushing, mountainous coastline of Alaska and Maine, the barrier island coasts facing the Atlantic, the steep, cliff-back headlands along the pacific coast, the marginal-sea type coastline of the Gulf region, and the coral reef coasts bordering Southern Florida and Hawaii.

Ice-pushing/mountainous coastline

These coastal regions along the northernmost part of the nation were affected predominantly by, along with the rest of the Pacific Coast, continuous tectonic activity, forming a very long, irregular, ridged, steep and mostly mountainous coastline. These environments are heavily occupied with permafrost and glaciers, which are the two major conditions affecting Alaska's Coastal Development.

Barrier island coastline

Barrier islands are a land form system that consists of fairly narrow strips of sand running parallel to the mainland and play a significant role in mitigating storm surges and oceans swells as natural storm events occur. The morphology of the various types and sizes of barrier islands depend on the wave energy, tidal range, basement controls, and sea level trends. The islands create multiple unique environments of wetland systems including marshes, estuaries, and lagoons.

Steep, cliff-backing abrasion coastline

The coastline along the western part of the nation consists of very steep, cliffed rock formations generally with vegetative slopes descending down and a fringing beach below.

The various sedimentary, metamorphic, and volcanic rock formations assembled along a tectonically disturbed environment, all with altering resistances running perpendicular, cause the ridged, extensive stretch of uplifted cliffs that form the peninsulas, lagoons, and valleys.

Marginal-sea type coastline

The southern banks of the United States border the Gulf of Mexico, intersecting numerous rivers, forming many inlets bays, and lagoons along its coast, consisting of vast areas of marsh and wetlands. This region of landform is prone to natural disasters yet highly and continuously developed, with man-made structures attaining to water flow and control.

Coral reef coastline

Coral reefs are located off the shores of the southern Florida and Hawaii consisting of rough and complex natural structures along the bottom of the ocean floor with extremely diverse ecosystems, absorbing up to ninety percent of the energy dissipated from wind-generated waves.

This process is a significant buffer for the inner-lying coastlines, naturally protecting and minimizing the impact of storm surge and direct wave damage. Because of the highly diverse ecosystems, these coral reefs not only provide for the shoreline protection, but also deliver an abundant amount of services to fisheries and tourism, increasing its economic value.

COASTAL GEOGRAPHY

The current trend does not mean that the Japanese automatically have a right to equidistance as the basis for these boundaries. Equidistance is certainly highly relevant and a basis for consideration, but, as the recent *Jan Mayen* and *St. Pierre and Miquelon* decisions illustrate, other considerations often result in delimitations different from equidistant lines. Thus, the proportionality analysis of the lengths of the relevant coastlines, the distance from the shoreline of the mainland and other considerations may cause the maritime boundary to be located closer to the Japanese islands than the equidistant line where these islands are small and far from the main Japanese islands, particularly where they face longer coastlines of other states.

The contemporary trend also means that claims based largely on natural prolongation, geology and geomorphology will not carry much weight. Equidistance, adjusted by coastline proportionality, and access to currently exploited resources will play important roles. Chinese claims based on the former considerations with regard to the South China Sea, the Yellow Sea and the East China Sea appear to have weakened as the law has evolved. Claims by Vietnam in the Gulf of Tonkin, and perhaps those of the Philippines and Vietnam in the South China Sea (especially regarding the Spratly Islands) may also have weakened.

In the Gulf of Tonkin, Vietnam seeks to limit the maritime boundary between it and China to a line allegedly established in 1887. This line runs due south from the ChineseVietnamese land boundary on the eastern side of the gulf, close to the large Chinese island of Hainan. It does not reflect the coastal front of Hainan and is located east of the equidistant line. The legal force of the 1887 line for maritime boundary delimitation purposes is highly questionable, but the matter must be resolved. If the 1887 line is not found to establish that boundary or historic rights, the geography of the coastline will become the significant factor under modern law. This argues for delimiting the boundary west of the 1887 line, reflecting the coastlines of the large Chinese island to the east and the Vietnamese mainland to the west. Such a boundary could approach equidistance.

In the Yellow Sea, maritime boundaries between the two Koreas and China have not been established. China has asserted as the basis for the maritime boundary in this area the limit of the silt deposited in the sea by the Yellow River. This silt line is located roughly two-thirds of the distance across the Yellow Sea towards the Korean Peninsula. China also claims fisheries zones in the Yellow Sea. While proof of historic exclusive fisheries may have some force in maritime boundary delimitation, the silt-line claim appears to be based on theories of natural prolongation and geomorphology. Such arguments have attracted little contemporary support, notwithstanding the language of the *North Sea Continental Shelf Cases*.

Moreover, delimitation based on coastal geography appears to be relatively simple, in the absence of complications arising from offshore islands distant from the mainland coasts or known significant geological or geomorphological features. The area may be characterised as essentially irregular upside-down Vwith equally long facing mainland coastlines. The maritime boundary is likely to approach equidistance if current practices are followed. The South Korean western islands (PaengnyOng-do and islands to the south) may create some complications in delimiting the maritime boundary between North and South Korea, a situation that would militate for some enclaving of the islands. A more complex picture is presented in the south, where offshore islands and disputed Chinese, Korean and Japanese claims come into play.

The serious territorial disputes over the islands in the South China Sea make it impossible to forecast the probable location of the maritime boundaries there. Certainly, sovereignty over the islands must be settled first. If one state prevails, the total effect of the islands may have a substantial impact on the overall location of the maritime boundaries. On the other hand, in the likely case that the islands are divided among several states, many complex minor maritime boundary lines will be required to divide the zones thus generated. Overall, however, the greater division of the South China Sea may be largely unaffected by the islands. The driving force will be the mainland coastlines.

The international maritime boundaries established by agreement and third-party dispute settlements demonstrate that small features distant from the mainland shore usually have a limited impact on the overall maritime delimitation within the 200-nautical-mile EEZ as a result of the application of various techniques for enclaving and discounting these features. Substantial islands far from shore may attract significant zones, but minor features may very well attract little area beyond the territorial sea. Many of the rocks, reefs, islets and islands in the areas discussed above are found in the latter category. Islands near the mainlands, however, can have augmenting effects. In general, the fundamental boundary delimitations for this area, if they are realised, will tend to enclave or ignore most of the islands and approach an equidistant line generated from the most substantial features. Many of the bordering states are opposite states, a circumstance that especially attracts resort to equidistance. Laterally related states will attract solutions that avoid cutoff effects (e.g., Brunei/ Malaysia).

Coastal geography appears to be the primary consideration in maritime delimitations. Connected to that consideration is a strong preference for delimiting maritime boundaries so that all the littoral states have access to the middle of the maritime areas in question. This approach was initially articulated by the first Judgement on maritime boundaries by the International Court of Justice in the *North Sea Continental Shelf Cases*. The Court found that a coastal state located within a concavity suffered a disadvantage from the convergence of equidistant lines that did not permit it to realise its equitable portion of the maritime zones involved. The result in that case was an agreement by the states on the North Sea to attribute areas to the state within the concavity that would give it access to the region towards the middle of the southern area of the North Sea. This regional view, based on equity, was even more explicitly endorsed by the tribunal in the *Guinea/Guinea-Bissau* arbitration, which took into account the entire regional configuration in order to produce an equitable result. The Court in the *Libya/Malta* case also considered the regional perspective in reaching its delimitation, as did the court of arbitration in the *Anglo-French* award. These cases establish that the regional perspective should be considered in maritime boundary delimitations.

This principle argues for providing each coastal state, where possible, with maritime boundaries that approach the high seas or the middle of the body of water in question. One of the interests to be protected is navigational and overflight rights, without the need to consider the interests or wishes of neighbouring states. In that regard, the North Sea solution gave Germany access towards the middle of the southern portion of the North Sea. In the *Anglo-French* case, this interest was considered to be generally relevant, but it was not a problem because of the favourable geographic circumstances.

In *St. Pierre and Miquelon,* the tribunal constructed a narrow 200-nautical-mile corridor that runs to the southernmost area within the tribunal's competence. This corridor can only be justified on the basis of providing access in the direction of the open sea. In the *Gulf of Fonseca* Judgement, the ICJ did not establish a corridor within the gulf, but, even more significantly, it confirmed Honduras's rights in the gulf including those of navigation and overflight.

In addition, it assured to Honduras the right to share in the area beyond the closing line of the gulf to the limits of national jurisdiction in the Pacific Ocean. A similar solution was employed by the ICJ in awarding the small island of Jan Mayen (Norway) a maritime boundary that approached the middle of the waters between it and Greenland (Denmark) despite Greenland's very long coastline in the area and Norway's access to the sea on the opposite side of Jan Mayen. These cases support the view that a state's access towards the open sea or the middle of the relevant body of water is a significant factor in delimiting maritime boundaries.

A review of the regional maps published in *International Maritime Boundaries* illustrates that in virtually all established maritime boundary delimitations (approximately 140), coastal states have not been cut off from access either to the open sea or to areas near the centre of the body of water on which their coastlines are located. This result is particularly apparent in the Caribbean Sea, the Persian Gulf, the Mediterranean Sea, the North Sea, the Baltic Sea, and the waters off Africa and South America. The African and South American situations are especially noteworthy because several of the coastal states agreed to maritime boundaries based on uniform cardinal directions and thus avoided cutoff effects on neighbouring states.

There are, however, a few exceptions to this practice. The maritime boundaries of Panama, albeit providing it with a fairly large maritime area, create a complete enclosure short of the middle of the bodies of water off both its coasts. In the Andaman Sea, the Indian islands of Andaman and Nicobar were given effect that substantially cut off similar access for Burma and Thailand. Nevertheless, these situations appear to be exceptions to the general practice in maritime boundary agreements.

ZONES NOT PERMITTED BY THE LOS CONVENTION

Settlement of the maritime boundaries in the East Asian area may have to take up the unusual maritime jurisdiction claims of the littoral states. The LOS Convention and the related general international law permit coastal states to claim internal or archipelagic waters delimited by the normal baselines and closing lines, historic bay closing lines, systems of straight baselines and archipelagic baselines; 12-nautical-mile territorial seas; 12-nautical-mile contiguous zones beyond the territorial sea; 200-nautical-mile EEZs measured from the baseline for the territorial sea; continental shelves; and — under certain circumstances not present here — continental shelves beyond the EEZ. No other zones of jurisdiction are permitted. The Convention purports to be comprehensive and may be considered general international law in this regard.

In the Central East Asian area, other maritime jurisdiction claims appear to be common. Thus, North Korea claims a special 50-nautical-mile military zone in the Sea of Japan drawn seaward from an unprecedented baseline not permitted by any of the law of the sea conventions. China claims certain military exclusion zones and special fishery zones in the East China Sea. China and South Korea claim that prior notification is required for the transit of foreign warships in their territorial seas despite the absence of any language in the articles on innocent passage and transit passage of the LOS Convention supporting such claims. Similarly, Japan claims to exclude vessels carrying nuclear weapons from transiting its territorial sea, notwithstanding these provisions.

Such special claims and zones certainly arose out of the controversial relations among the neighbouring states in the region. In their bilateral relations, however, the coastal states may find that these unusual, if not unsanctioned, claims under the general international law of the sea do not present obstacles to maritime boundary delimitation. None of these claims violate jus cogens norms. Thus, these states may enter into international agreements that establish maritime boundaries accepting these unusual claims in their bilateral relations. For example, the USSR was apparently willing to acquiesce in the unprecedented claims of North Korea when it agreed to their maritime boundaries in the area.

The quid pro quo may have been North Korean acquiescence in the Soviet historic bay claim over Peter the Great Bay. Other unusual claims may pertain more directly to disputed boundaries, such as the Chinese special zones in the East China Sea, which appear to run into the eastern portions of the sea claimed by Japan and South Korea. The establishment of a definitive traditional maritime boundary would require the resolution of these Chinese claims, although joint arrangements or special rights might be included in the settlement. Of course, such settlements may not affect third-state navigation and overflight rights and the underlying disputes may therefore have to be resolved by other means.

FUTURE SETTLEMENTS

The entry into force of the Law of the Sea Convention may encourage the states in the region to settle their maritime boundaries. The Convention provides the requisite codified law and thus eliminates some grounds for dispute. It also clarifies some new law that bears on maritime boundary delimitation. Furthermore, it establishes a comprehensive system for resolving disputes involving the law of the sea.

This system permits states to invoke a variety of techniques, including negotiation, conciliation, arbitration and adjudication, to arrive at boundary settlements. The ready availability of conciliation procedures and the participation of specialists may encourage Central East Asian states to accept delimitation as well as other solutions. Since these states generally do not favour litigation, good offices and conciliation may play important roles in resolving.

The Convention does provide for compulsory binding settlement of disputes arising out of the application or interpretation of its provisions. The choice of procedure is left to the states concerned and, in the case of maritime boundary disputes, permits states to exclude certain compulsory and binding dispute settlement procedures. This is complex and somewhat ambiguous. It is clear, however, that unless a state submits a contrary declaration in writing at the time of signature, ratification or accession, maritime boundary disputes involving that state will be subject to the compulsory binding dispute settlement procedures established by the Convention. If such a declaration is deposited, it will not fully exclude disputes arising subsequent to entry into force of the Convention (November 16, 1994) from the dispute settlement system when negotiations have been exhausted. Those new disputes will remain subject to compulsory conciliation.

Furthermore, the Convention states that "such submission" excludes disputes "concerning sovereignty or other rights over continental or insular land territory." "Such submission" could be interpreted broadly to refer to all compulsory binding dispute settlement procedures regarding maritime boundaries. But its meaning is narrower. It appears in the context of a unilateral declaration excluding maritime boundary disputes from the normal dispute settlement provisions of the Convention (Part XV), and it follows directly upon language regarding submission to conciliation of newly arising disputes.

Thus, the text should be interpreted to mean that this special conciliation procedure may not be used for land territory issues. Absent a declaration, however, boundary disputes are properly subject to compulsory binding dispute settlement and the other dispute settlement procedures of the Convention. The relevant forum is not limited by this from addressing all issues necessary to determine the maritime boundary, including territorial sovereignty if it is germane. Furthermore, of course, states may specifically agree to submit a maritime boundary dispute to a different third-party forum for resolution, with or without qualifications.

It remains an open question whether any of the disputes in the Central East Asian area will be resolved through binding third-party dispute settlement. Conciliation under the LOS Convention or outside the Convention may be the preferred route if negotiations fail. Conciliation may produce results that approximate a binding settlement since the process and consequent report may put the disputing states under strong mutual and international pressure to accept those results or otherwise settle the matter. On the other hand, if a state elects to exclude maritime boundary disputes from compulsory binding settlement, compulsory conciliation applies only to newly arising maritime boundary disputes, which appears to rule out all of the disputes discussed in this chapter.

Since World War II, relations between certain of the littoral states in the region have been difficult. These difficulties have lessened the prospects for settlement of the maritime boundary disputes. Thus, North and South Korea have fought a war and enmity remains. China and Taiwan have entertained similarly difficult relations. Many nations in the region consider Japan to have been the aggressor during World War II and treat it warily. Their caution has increased as Japan has regained its economic and military strength. Vietnam was divided for a long time, and war consumed the energies of its two parts. Finally, some of these states feared that the Soviet Union harboured aggressive tendencies in the region.

This situation may now be changing as a result of the demise of the USSR and Russia's assumption of the Soviet territories in Central East Asia. In addition, progress is being made in bringing North Korea out of isolation, and a rapprochement between North and South Korea is a possibility. China and Taiwan have recently exchanged diplomatic visits and relations may move towards regularisation. Russia is not viewed as having maintained the Soviet aggressive tendencies and settlement of the dispute over the sovereignty of the Kuril Islands is under consideration. China has become increasingly integrated into the international community, but it has also begun to project strong regional aspirations. Many of these developments may bode well for discussions and, perhaps, ultimate resolution of the maritime boundary disputes.

For several years Indonesia has led efforts to settle the disputes regarding the South China Sea without resolving the underlying territorial questions or establishing maritime boundaries. Non-boundary-based solutions may be the most promising in light of the difficult sovereignty and boundary issues involved. The early resolution of these disputes may succeed in preventing costly international crises. The South China Sea is a major shipping route, especially for supertankers moving from the Persian Gulf to Japan and Korea, container ships sailing between Hong Kong and Singapore, and military vessels. The extreme weather conditions, as well as the many shoals, pose a constant risk of major pollution incidents. Other environmental problems are also likely. Fishing by many nations in the area carries the risk of overexploitation and collapse of the fisheries.

Regional regimes to address these issues without prejudice to the underlying claims could be created. A variety of regional authorities could address these regional issues. They might vary with regard to the degree and nature of decision making, government control and enforcement roles. A possible first step might be to develop plans to respond effectively to potential environmental disasters. These arrangements might be addressed to common concerns and would be least likely to raise jurisdictional issues. They might even help to build trust among the states in the region. A second step might be to develop regional environmental standards implementing the pertinent provisions of the LOS Convention to assure that generally accepted international minimum standards are fully observed.

Higher environmental standards consistent with the Convention might be established within the littoral states' EEZs and for their flag vessels operating in the region, as well as with regard to the ocean dumping of materials shipped from their ports. Special standards might be adopted for particularly delicate areas of the EEZ, as permitted under the Convention. Once mutual cooperation on these matters is established, the region might move to a third step addressing fisheries management from a regional perspective by promoting a coordinated policy of conservation and effective enforcement procedures. The norms of the LOS Convention, the fact that these are sernienclosed seas and the impending United Nations Convention on Straddling Fish Stocks and Highly Migratory Fish Stocks should help to guide this effort.

Finally, cooperative arrangements for the development of seabed mineral resources could be addressed. Since all of these arrangements could be established without prejudice to the underlying territorial and maritime boundary issues, they might constitute optimal solutions that would defer the more politically charged issues to the indefinite future. Although the informal discussions promoted by Indonesia may be productive, recent reports are not encouraging. For a cooperative non-boundary solution to succeed, the participating states must have good relations that facilitate the mutual confidence necessary for effective working relationships. The current relations in the region do not bode well in this regard, but recent political developments may change this prognosis, and a phased process might help to eliminate the barriers.

The littoral states may have other incentives to reach settlements, especially before strong conflicts arise or crises occur. Recent maritime boundary practices and decisions tend to avoid upsetting long-standing resource development operations. Where many states are exploiting the living resources in an area, the solution may be to divide the resource among them. Time is on the side of the state pursuing development regardless of historic or traditional activities. If economic development of a marine resource takes place, it will be difficult to dislodge that state completely from the area by negotiation or third-party dispute settlement. Thus, the status quo may work against some of the states involved. They might wish to obtain boundary settlements before other states establish a new status quo of resource development. Recently, states in the region have unilaterally attempted to establish a new status quo that would be difficult to upset. Especially in the South China Sea, hydrocarbon leases have been issued, military garrisons have been built on the Spratlys, and military skirmishes have been fought.

If, ultimately, some or all of the disputes in the area are to be settled by the establishment of maritime boundaries (by agreement or third-party settlement), current developments suggest that certain outcomes are likely. First, China will probably not establish a right to the whole of the South China Sea. Second, no state will be able to draw baselines that encompass large areas of the waters in question by connecting islands on the basis of archipelagic baseline or straight baseline system claims. Third, many of the features in the region will be found to be "rocks which cannot sustain human habitation," and thus will not be entitled independently to zones beyond the territorial sea and perhaps the contiguous zone. Fourth, many of the features that are not such rocks will be considered too small to affect the overall location of the maritime boundary and, instead, will be enclaved or otherwise provide the sovereign with limited maritime areas.

Fifth, geology and geomorphology will have little impact on the delimitations. Sixth, the maritime boundary delimitations will rather be based primarily on the geography of the littoral states' mainland coastlines and the larger islands. Since, viewed regionally, most of the states' boundaries are opposite, equidistance will play an important role in the initial positioning of these boundaries. But they also will be influenced by the coastline lengths, resource development activities and, perhaps, other functional considerations. Finally, the entire area will be included in the EEZ of the littoral states while such traditional high seas Freedoms as navigation and overflight will be protected. Under this regime, the coastal states have certain rights, including the right to exploit the living marine resources and the exclusive right to exploit the non-living resources of the seabed and subsoil, as well as duties to protect the environment and manage the living and. non-living resources.

The result may be a complicated system of maritime boundaries. Since the ecosystem and the resources would not respect these boundaries, the littoral states may be motivated to adopt more cooperative regimes, notwithstanding the possible establishment of maritime boundaries and the settlement of the sovereignty disputes. Such settlements, if they ever take place, will take time. Whether they will lead to cooperative arrangements remains uncertain. In light of the complexity of the situation in the Central East Asian maritime area, the littoral states may prefer to settle the territorial disputes and the maritime boundaries before moving towards substantial regional or bilateral cooperative arrangements. But the intractability of some of the problems may make regional solutions attractive, by enabling the interested states to resolve pressing issues of environmental protection and resource development without reaching the underlying sovereignty and boundary issues. Now that the LOS Convention is in force, the rules are more settled and a framework for the resolution of such issues is available, which may encourage states to settle their disputes voluntarily. However, the tensions between the littoral states make any solution to these complicated issues in the near future problematic.

5

GIS Techniques and Technology

Modern GIS technologies use digital information, for which various digitized data creation methods are used. The most common method of data creation is digitization, where a hard copy map or survey plan is transferred into a digital medium through the use of a computer-aided design programme, and geo-referencing capabilities.

With the wide availability of ortho-rectified imagery, heads-up digitizing is becoming the main avenue through which geographic data is extracted. Heads-up digitizing involves the tracing of geographic data directly on top of the aerial imagery instead of by the traditional method of tracing the geographic form on a separate digitizing tablet.

Relating Information from Different Sources

GIS uses spatio-temporal location as the key index variable for all other information. Just as a relational database containing text or numbers can relate many different tables using common key index variables, GIS can relate otherwise unrelated information by using location as the key index variable. The key is the location and/or extent in space-time. Any variable that can be located spatially, and increasingly also temporally, can be referenced using a GIS. Locations or extents in Earth space-time may be recorded as dates/times of occurrence, and x, y, and z coordinates representing, longitude, latitude, and elevation, respectively. These GIS coordinates may represent other quantified systems of temporo-spatial reference. Units applied to recorded temporal-spatial data can vary widely but all Earth-based spatial-temporal location and extent references should, ideally, be relatable to one another and ultimately to a "real" physical location or extent in space-time.

Related by accurate spatial information, an incredible variety of real-world and projected past or future data can be analysed, interpreted and represented to facilitate education and decision making. This key characteristic of GIS has begun to open new avenues of scientific inquiry into behaviours and patterns of previously considered unrelated real-world information.

GIS THROUGH HISTORY

Some 35,000 years ago, Cro-Magnon hunters drew pictures of the animals they hunted on the walls of caves near Lascaux, France. Associated with the animal drawings are track lines and tallies thought to depict migration routes. These early records followed the two-element structure of modern geographic information systems (GIS): a graphic file linked to an attribute database.

Today, biologists use collar transmitters and satellite receivers to track the migration routes of caribou and polar bears to help design programmes to protect the animals. In a GIS, the migration routes were indicated by different colours for each month for 21 months. Researchers then used the GIS to superimpose the migration routes on maps of oil development plans to determine the potential for interference with the animals.

Mapmaking

Researchers are working to incorporate the mapmaking processes of traditional cartographers into GIS technology for the automated production of maps. One of the most common products of a GIS is a map. Maps are generally easy to make using a GIS and they are often the most effective means of communicating the results of the GIS process. Therefore, the GIS is usually a prolific producer of maps. The users of a GIS must be concerned with the quality of the maps produced because the GIS normally does not regulate common cartographic principles. One of these principles is the concept of generalization, which deals with the content and detail of information at various scales. The GIS user can change scale at the push of a button, but controlling content and detail is often not so easy. Mapmakers have long recognized that content and detail need to change as the scale of the map changes. For example, the State of New Jersey can be mapped at various scales, from the small scale of 1:500,000 to the larger scale of 1:250,000 and the yet larger scale of 1:100,000,but each scale requires an appropriate level of generalization.

Site Selection

The U.S. Geological Survey (USGS), in a cooperative project with the Connecticut Department of Natural Resources, digitized more than 40 map layers for the areas covered by the USGS Broad Brook and Ellington 7.5-minute topographic quadrangle maps. This information can be combined and manipulated in a GIS to address planning and natural resource issues. GIS information was used to locate a potential site for a new water well within half a mile of the Somers Water Company service area. To prepare the analysis, cartographers stored digital maps of the water service areas in the GIS. They used the proximity function in the GIS to draw a half-mile buffer zone around the water company service area. This buffer zone was the "window" used to view and combine the various map coverages relevant to the well site selection.

The land use and land cover map for the two areas shows that the area is partly developed. A GIS was used to select undeveloped areas from the land use and land cover map as the first step in finding well sites. The developed areas were eliminated from further consideration. The quality of water in Connecticut streams is closely monitored. Some of the streams in the study area were known to be unusable as drinking water sources. To avoid pulling water from these streams into the wells, 100-meter buffer zones were created around the unsuitable streams using the GIS, and the zones were plotted on the map. The areas in blue have the characteristics desired for a water well site.

Point sources of pollution are recorded by the Connecticut Department of Natural Resources. These records consist of a location and a text description of the pollutant. To avoid these toxic areas, a buffer zone of 500 meters was established around each point. This information was combined with the previous two map layers to produce a new map of areas suitable for well sites. Points sources of pollution in the water service area are identified and entered into a GIS.

The map of surficial geology shows the earth materials that lie above bedrock. Since the area under consideration in Connecticut is covered by glacial deposits, the surface consists largely of sand and gravel, with some glacial till and fine-grained sediments. Of these materials, sand and gravel are the most likely to store water that could be tapped with wells. Areas underlain by sand and gravel were selected from the surficial geology map. They were combined with the results of the previous selections to produce a map consisting of: (1) sites in underdeveloped areas underlain by sand and gravel, (2) more than 500 meters from point sources of pollution, and (3) more than 100 meters from unsuitable streams. A map that shows the thickness of saturated sediments was created by using the GIS to subtract the bedrock elevation from the surface elevation. For this analysis, areas having more than 40 feet of saturated sediments were selected and combined with the previous overlays.

The resulting site selection areas that are undeveloped, are situated outside the buffered pollution areas, and are underlain by 40 feet or more of water-saturated sand and gravel. Because of map resolution and the limits of precision in digitizing, the very small polygons (areas) may not have all of the characteristics analysed, so another GIS function was used to screen out areas smaller than 10 acres. The final six sites are displayed with the road and stream network and selected place names for use in the field. Potential water well sites, roads, streams and place names. The process illustrated by this site selection analysis has been used for many common applications, including transportation planning and waste disposal site location. The technique is particularly useful when several physical factors must be considered and integrated over a large area.

Emergency Response Planning

A GIS was used to combine road network and earth science information to analyse the effect of an earthquake on the response time of fire and rescue squads. The area covered by the USGS Sugar House 7.5-minute topographic quadrangle map was selected for the study because it includes both undeveloped areas in the mountains and a part of Salt Lake City. Detailed earth science information was available for the entire region. The road network from a USGS digital line graph includes information on the types of roads, which range from rough trails to divided highways. The locations of fire stations were plotted on the road network. A GIS function called network analysis was used to calculate the time necessary for emergency vehicles to travel from the fire stations to different areas of the city. The network analysis function considers two elements: (1) distance from the fire station, and (2) speed of travel based on the type of road. The analysis shows that under normal conditions, most of the area within the city will be served in less than 7 minutes and 30 seconds because of the distribution and density of fire stations and the continuous network of roads.

The accompanying illustration depicts the blockage of the road network that would result from an earthquake, assuming that any road crossing the fault trace would become impassable. The primary effect on emergency response time would occur in neighbourhoods west of the fault trace, where travel times from the fire stations would be noticeably lengthened. Road network of area covered by the Sugar House quadrangle plotted from USGS digital line graph data, indicating the locations of fire stations and travel times of emergency vehicles.

Areas in blue can receive service within 2½minutes, area in green within 5 minutes, areas in yellow within 7½ minutes, and areas in magenta within 10 minutes. Areas in white cannot receive service within 10 minutes. After faulting, initial model. Network analysis in a GIS produces a map of travel times from the stations after faulting. The fault is in red. Emergency response times have increased for areas west of the fault. The Salt Lake City area lies on lake sediments of varying thicknesses.

These sediments range from clay to sand and gravel, and most are water-saturated. In an earthquake, these materials may momentarily lose their ability to support surface structures, including roads.

The potential for this phenomenon, known as liquefaction, is shown in a composite map portraying the inferred relative stability of the land surface during an earthquake. Areas near the fault and underlain by thick, loosely consolidated, water-saturated sediments will suffer the most intense surface motion during an earthquake. Areas on the mountain front with thin surface sediments will experience less additional ground acceleration. The map of liquefaction potential was combined with the road network analysis to show the additional effect of liquefaction on response times.

The areas near the fault, as well as those underlain by thick, water-saturated sediments, are subject to more road disruptions and slower emergency response than are other areas of the city. Map of potential ground l liquefaction during an earthquake.

The least stable areas are shown by yellows and oranges, the most stable by grays and browns. After faulting, final model. An earthquake on emergency travel times is reduced by combining the liquefaction potential information.

GIS: COMPUTER SYSTEM OF CAPTURING, STORING AND ANALYSING

A GIS is a computer system capable of capturing, storing, analysing, and displaying geographically referenced information; that is, data identified according to location. Practitioners also define a GIS as including the procedures, operating personnel, and spatial data that go into the system.

The power of a GIS comes from the ability to relate different information in a spatial context and to reach a conclusion about this relationship. Most of the information we have about our world contains a location reference, placing that information at some point on the globe.

When rainfall information is collected, it is important to know where the rainfall is located. This is done by using a location reference system, such as longitude and latitude, and perhaps elevation. Comparing the rainfall information with other information, such as the location of marshes across the landscape, may show that certain marshes receive little rainfall. This fact may indicate that these marshes are likely to dry up, and this inference can help us make the most appropriate decisions about how humans should interact with the marsh. A GIS, therefore, can reveal important new information that leads to better decisionmaking.

Many computer databases that can be directly entered into a GIS are being produced by Federal, State, tribal, and local governments, private companies, academia, and nonprofit organizations. Different kinds of data in map form can be entered into a GIS. A GIS can also convert existing digital information, which may not yet be in map form, into forms it can recognize and use. For example, digital satellite images can be analysed to produce a map of digital information about land use and land cover. Likewise, census or hydrologic tabular data can be converted to a maplike form and serve as layers of thematic information in a GIS.

Data Capture

How can a GIS use the information in a map? If the data to be used are not already in digital form, that is, in a form the computer can recognize, various techniques can capture the information. Maps can be digitized by hand-tracing with a computer mouse on the screen or on a digitizing tablet to collect the coordinates of features. Electronic scanners can also convert maps to digits. Cordinates from Global Positioning System (GPS) receivers can also be uploaded into a GIS. A GIS can be used to emphasize the spatial relationships among the objects being mapped. While a computer-aided mapping system may represent a road simply as a line, a GIS may also recognize that road as the boundary between wetland and urban development between two census statistical areas.

Data capture—putting the information into the system—involves identifying the objects on the map, their absolute location on the Earth's surface, and their spatial relationships. Software tools that automatically extract features from satellite images or aerial photographs are gradually replacing what has traditionally been a time-consuming capture process. Objects are identified in a series of attribute tables—the "information" part of a GIS. Spatial relationships, such as whether features intersect or whether they are adjacent, are the key to all GIS-based analysis.

Data Integration

A GIS makes it possible to link, or integrate, information that is difficult to associate through any other means. Thus, a GIS can use combinations of mapped variables to build and analyse new variables. Data integration is the linking of information in different forms through a GIS. For example, using GIS technology, it is possible to combine agricultural records with hydrography data to determine which streams will carry certain levels of fertilizer runoff. Agricultural records can indicate how much pesticide has been applied to a parcel of land. By locating these parcels and intersecting them with streams, the GIS can be used to predict the amount of nutrient runoff in each stream. Then as streams converge, the total loads can be calculated downstream where the stream enters a lake.

Projection and Registration

A property ownership map might be at a different scale than a soils map. Map information in a GIS must be manipulated so that it registers, or fits, with information gathered from other maps. Before the digital data can be analysed, they may have to undergo other manipulations—*projection conversions,* for example—that integrate them into a GIS. Projection is a fundamental component of mapmaking. A projection is a mathematical means of transferring information from the Earth's three-dimensional, curved surface to a two-dimensional medium—paper or a computer screen. Different projections are used for different types of maps because each projection is particularly appropriate for certain uses. For example, a projection that accurately represents the shapes of the continents will distort their relative sizes.

Since much of the information in a GIS comes from existing maps, a GIS uses the processing power of the computer to transform digital information, gathered from sources with different projections, to a common projection. An elevation image classified from a satellite image of Minnesota exists in a different scale and projection than the lines on the digital file of the State and province boundaries. The elevation image has been reprojected to match the projection and scale of the State and province boundaries.

Data Structures

Can a land use map be related to a satellite image, a timely indicator of land use? Yes, but because digital data are collected and stored in different ways, the two data sources may not be entirely compatible. Therefore, a GIS must be able to convert data from one structure to another. Satellite image data that have been interpreted by a computer to produce a land use map can be "read into" the GIS in raster format. Raster data files consist of rows of uniform cells coded according to data values. An example is land cover classification. Raster files can be manipulated quickly by the computer, but they are often less detailed and may be less visually appealing than vector data files, which can approximate the appearance of more traditional hand-drafted maps. Vector digital data have been captured as points, lines (a series of point coordinates), or areas (shapes bounded by lines). An example of data typically held in a vector file would be the property boundaries for a particular housing subdivision.

Data Modelling

It is impossible to collect data over every square meter of the Earth's surface. Therefore, samples must be taken at discrete locations. A GIS can be used to depict two-and three-dimensional characteristics of the Earth's surface, subsurface, and atmosphere from points where samples have been collected. For example, a GIS can quickly generate a map with isolines that indicate the pH of soil from test points. Such a map can be thought of as a soil pH contour map. Many sophisticated methods can estimate the characteristics of surfaces from a limited number of point measurements. Two-and three-dimensional contour maps created from the surface modelling of sample points from pH measurements can be analysed together with any other map in a GIS covering the area.

The way maps and other data have been stored or filed as layers of information in a GIS makes it possible to perform complex analyses. A crosshair pointer (top) can be used to point at a location stored in a GIS. The bottom illustration depicts a computer screen containing the kind of information stored about the location—for example, the latitude, longitude, projection, coordinates, closeness to wells, sources of production, roads, and slopes of land.

Information Retrieval

What do you know about the swampy area at the end of your street? With a GIS you can "point" at a location, object, or area on the screen and retrieve recorded information about it from offscreen files. Using scanned aerial photographs as a visual guide, you can ask a GIS about the geology or hydrology of the area or even about how close a swamp is to the end of a street. This type of analysis allows you to draw conclusions about the swamp's environmental sensitivity.

Topological Modelling

Have there ever been gas stations or factories that operated next to the swamp? Were any of these uphill from and within 2 miles of the swamp? A GIS can recognize and analyse the spatial relationships among mapped phenomena. Conditions of adjacency (what is next to what), containment (what is enclosed by what), and proximity (how close something is to something else) can be determined with a GIS.

Networks

When nutrients from farmland are running off into streams, it is important to know in which direction the streams flow and which streams empty into other streams. This is done by using a linear network. It allows the computer to determine how the nutrients are transported downstream. Additional information on water volume and speed throughout the spatial network can help the GIS determine how long it will take the nutrients to travel downstream. A GIS can simulate the movement of materials along a network of lines. These illustrations show the route of pollutants through a stream system. Flow directions are indicated by arrows.

Overlay

Using maps of wetlands, slopes, streams, land use, and soils, the GIS might produce a new map layer or overlay that ranks the wetlands according to their relative sensitivity to damage from nutrient runoff.

Data Output

A critical component of a GIS is its ability to produce graphics on the screen or on paper to convey the results of analyses to the people who make decisions about resources. Wall maps, Internet-ready maps, interactive maps, and other graphics can be generated, allowing the decisionmakers to visualize and thereby understand the results of analyses or simulations of potential events.

Framework for Cooperation

The use of a GIS can encourage cooperation and communication among the organizations involved in environmental protection, planning, and resource management. The collection of data for a GIS is costly. Data collection can require very specialized computer equipment and technical expertise. Standard data formats ease the exchange of digital information among users of different systems. Standardization helps to stretch data collection funds further by allowing data sharing, and, in many cases, gives users access to data that they could not otherwise collect for economic or technical reasons. Organizations such as the University Consortium for Geographic Information Science and the Federal Geographic Data Committee seek to encourage standardization efforts.

GIS UNCERTAINTIES

GIS accuracy depends upon source data, and how it is encoded to be data referenced. Land Surveyors have been able to provide a high level of positional accuracy utilizing the GPS derived positions. The high-resolution digital terrain and aerial imagery, the powerful computers, Web technology, are changing the quality, utility, and expectations of GIS to serve society on a grand scale, but nevertheless there are other source data that has an impact on the overall GIS accuracy like: paper maps that are not found to be very suitable to achieve the desired accuracy since the aging of maps affects their dimensional stability. In developing a Digital Topographic Data Base for a GIS, topographical maps are the main source of data.

Aerial photography and satellite images are extra sources for collecting data and identifying attributes which can be mapped in layers over a location facsimile of scale. The scale of a map and geographical rendering area representation type are a very important aspects since the information content depends mainly on the scale set and resulting locatability of the map's representations.

In order to digitize a map, the map has to be checked within theoretical dimensions, then scanned into a raster format, and resulting raster data has to be given a theoretical dimension by a rubber sheeting/warping technology process. Uncertainty is a significant problem in designing a GIS because spatial data tend to be used for purposes for which they were never intended. Some maps were made many decades ago, where at that time the computer industry was not even in its perspective establishments. This has led to historical reference maps without common norms. Map accuracy is a relative issue of minor importance in cartography. All maps are established for communication ends. Maps use a historically constrained technology of pen and paper to communicate a view of the world to their users. Cartographers feel little need to communicate information based on accuracy, for when the same map is digitized and input into a GIS, the mode of use often changes. The new uses extend well beyond a determined domain for which the original map was intended and designed. A quantitative analysis of maps brings accuracy issues into focus. The electronic and other equipment used to make measurements for GIS is far more precise than the machines of conventional map analysis. The truth is that all geographical data are inherently inaccurate, and these inaccuracies will propagate through GIS operations in ways that are difficult to predict, yet have goals of conveyance in mind for original design.

Accuracy Standards for 1:24000 Scales Map: 1:24,000 ± 40.00 feet This means that when we see a point or attribute on a map, its "probable" location is within a +/- 40 foot area of its rendered reference, just as to area representations and scale. A GIS can also convert existing digital information, which may not yet be in map form, into forms it can recognize, employ for its data analysis processes, and use in forming mapping output. For example, digital satellite images generated through remote sensing can be analysed to produce a map-like layer of digital information about vegetative covers on land locations. Another fairly recently developed resource for naming GIS location objects is the Getty Thesaurus of Geographic Names which is a structured vocabulary containing about 1,000,000 names and other information about places.Likewise, researched census or hydrological tabular data can be displayed in map-like form, serving as layers of thematic information for forming a GIS map.

Data Representation

GIS data represents real objects with digital data determining the mix. Real objects can be divided into two abstractions: discrete objects and continuous fields. Traditionally, there are two broad methods used to store data in a GIS for both kinds of abstractions mapping references: raster images and vector. Points, lines, and polygons are the stuff of mapped location attribute references. A new hybrid method of storing data is that of identifying point clouds, which combine three-dimensional points with RGB information at each point, returning a "3D colour image". GIS Thematic maps then are becoming more and more realistically visually descriptive of what they set out to show or determine.

Raster

A raster data type is, in essence, any type of digital image represented by reducible and enlargeable grids. Anyone who is familiar with digital photography will recognize the Raster graphics pixel as the smallest individual grid unit building block of an image, usually not readily identified as an artifact shape until an image is produced on a very large scale. A combination of the pixels making up an image colour formation scheme will compose details of an image, as is distinct from the commonly used points, lines, and polygon area location symbols of scalable vector graphics as the basis of the vector model of area attribute rendering.

While a digital image is concerned with its output blending together its grid based details as an identifiable representation of reality, in a photograph or art image transferred into a computer, the raster data type will reflect a digitized abstraction of reality dealt with by grid populating tones or objects, quantities, cojoined or open boundaries, and map relief schemas. Aerial photos are one commonly used form of raster data, with one primary purpose in mind: to display a detailed image on a map area, or for the purposes of rendering its identifiable objects by digitization. Additional raster data sets used by a GIS will contain information regarding elevation, a digital elevation model, or reflectance of a particular wavelength of light, Landsat, or other electromagnetic spectrum indicators. Raster data type consists of rows and columns of cells, with each cell storing a single value.

Raster data can be images with each pixel containing a colour value. Additional values recorded for each cell may be a discrete value, such as land use, a continuous value, such as temperature, or a null value if no data is available. While a raster cell stores a single value, it can be extended by using raster bands to represent RGB colours, colourmaps or an extended attribute table with one row for each unique cell value.

The resolution of the raster data set is its cell width in ground units. Raster data is stored in various formats; from a standard file-based structure of TIF, JPEG, etc. to binary large object data stored directly in a relational database management system similar to other vector-based feature classes. Database storage, when properly indexed, typically allows for quicker retrieval of the raster data but can require storage of millions of significantly sized records.

Vector

In a GIS, geographical features are often expressed as vectors, by considering those features as geometrical shapes.

Different geographical features are expressed by different types of geometry:

- *Points*: Zero-dimensional points are used for geographical features that can best be expressed by a single point reference—in other words, by simple location. Examples include wells, peaks, features of interest, and trailheads. Points convey the least amount of information of these file types. Points can also be used to represent areas when displayed at a small scale. For example, cities on a map of the world might be represented by points rather than polygons. No measurements are possible with point features.
- *Lines or polylines*: One-dimensional lines or polylines are used for linear features such as rivers, roads, railroads, trails, and topographic lines. Again, as with point features, linear features displayed at a small scale will be represented as linear features rather than as a polygon. Line features can measure distance.
- *Polygons*: Two-dimensional polygons are used for geographical features that cover a particular area of the earth's surface. Such features may include lakes, park boundaries, buildings, city boundaries, or land uses. Polygons convey the most amount of information of the file types. Polygon features can measure perimeter and area.

Each of these geometries are linked to a row in a database that describes their attributes. For example, a database that describes lakes may contain a lake's depth, water quality, pollution level.

This information can be used to make a map to describe a particular attribute of the dataset. For example, lakes could be coloured depending on level of pollution. Different geometries can also be compared. For example, the GIS could be used to identify all wells that are within one kilometre of a lake that has a high level of pollution.

Vector features can be made to respect spatial integrity through the application of topology rules such as 'polygons must not overlap'. Vector data can also be used to represent continuously varying phenomena. Contour lines and triangulated irregular networks are used to represent elevation or other continuously changing values. TINs record values at point locations, which are connected by lines to form an irregular mesh of triangles. The face of the triangles represent the terrain surface.

Advantages and Disadvantages

There are some important advantages and disadvantages to using a raster or vector data model to represent reality:

- Raster datasets record a value for all points in the area covered which may require more storage space than representing data in a vector format that can store data only where needed.

- Raster data allows easy implementation of overlay operations, which are more difficult with vector data.
- Vector data can be displayed as vector graphics used on traditional maps, whereas raster data will appear as an image that may have a blocky appearance for object boundaries.
- Vector data can be easier to register, scale, and re-project, which can simplify combining vector layers from different sources.
- Vector data is more compatible with relational database environments, where they can be part of a relational table as a normal column and processed using a multitude of operators.
- Vector file sizes are usually smaller than raster data, which can be 10 to 100 times larger than vector data.
- Vector data is simpler to update and maintain, whereas a raster image will have to be completely reproduced.
- Vector data allows much more analysis capability, especially for "networks" such as roads, power, rail, telecommunications, etc. Raster data will not have all the characteristics of the features it displays.

Non-spatial Data

Additional non-spatial data can also be stored along with the spatial data represented by the coordinates of a vector geometry or the position of a raster cell. In vector data, the additional data contains attributes of the feature. For example, a forest inventory polygon may also have an identifier value and information about tree species. In raster data the cell value can store attribute information, but it can also be used as an identifier that can relate to records in another table.

Software is currently being developed to support spatial and non-spatial decision-making, with the solutions to spatial problems being integrated with solutions to non-spatial problems. The end result with these Flexible Spatial Decision-Making Support Systems is expected to be that non-experts will be able to use GIS, along with spatial criteria, and simply integrate their non-spatial criteria to view solutions to multi-criteria problems. This system is intended to assist decision-making.

Data Capture

Data capture—entering information into the system—consumes much of the time of GIS practitioners. There are a variety of methods used to enter data into a GIS where it is stored in a digital format. Existing data printed on paper or PET film maps can be digitized or scanned to produce digital data. A digitizer produces vector data as an operator traces points, lines, and polygon boundaries from a map. Scanning a map results in raster data that could be further processed to produce vector data. Survey data can be directly entered into a GIS from digital data collection systems on survey instruments using a technique called Coordinate Geometry.

Positions from a Global Navigation Satellite System like Global Positioning System, another survey tool, can also be directly entered into a GIS. Current trend is data collection and field mapping carried out directly with field computers. New technologies allow to create maps as well as analysis directly in the field, projects are more efficient and mapping is more accurate. Remotely sensed data also plays an important role in data collection and consist of sensors attached to a platform. Sensors include cameras, digital scanners and LIDAR, while platforms usually consist of aircraft and satellites.

The majority of digital data currently comes from photo interpretation of aerial photographs. Soft copy workstations are used to digitize features directly from stereo pairs of digital photographs. These systems allow data to be captured in two and three dimensions, with elevations measured directly from a stereo pair using principles of photogrammetry. Currently, analog aerial photos are scanned before being entered into a soft copy system, but as high quality digital cameras become cheaper this step will be skipped.

Satellite remote sensing provides another important source of spatial data. Here satellites use different sensor packages to passively measure the reflectance from parts of the electromagnetic spectrum or radio waves that were sent out from an active sensor such as radar. Remote sensing collects raster data that can be further processed using different bands to identify objects and classes of interest, such as land cover.

When data is captured, the user should consider if the data should be captured with either a relative accuracy or absolute accuracy, since this could not only influence how information will be interpreted but also the cost of data capture. In addition to collecting and entering spatial data, attribute data is also entered into a GIS. For vector data, this includes additional information about the objects represented in the system. After entering data into a GIS, the data usually requires editing, to remove errors, or further processing.

For vector data it must be made "topologically correct" before it can be used for some advanced analysis. For example, in a road network, lines must connect with nodes at an intersection. Errors such as undershoots and overshoots must also be removed. For scanned maps, blemishes on the source map may need to be removed from the resulting raster. For example, a fleck of dirt might connect two lines that should not be connected.

Raster-to-vector Translation

Data restructuring can be performed by a GIS to convert data into different formats. For example, a GIS may be used to convert a satellite image map to a vector structure by generating lines around all cells with the same classification, while determining the cell spatial relationships, such as adjacency or inclusion.

More advanced data processing can occur with image processing, a technique developed in the late 1960s by NASA and the private sector to provide contrast enhancement, false colour rendering and a variety of other techniques including use of two dimensional Fourier transforms. Since digital data is collected and stored in various ways, the two data sources may not be entirely compatible. So a GIS must be able to convert geographic data from one structure to another.

Projections, Coordinate Systems and Registration

A property ownership map and a soils map might show data at different scales. Map information in a GIS must be manipulated so that it registers, or fits, with information gathered from other maps. Before the digital data can be analysed, they may have to undergo other manipulations—projection and coordinate conversions, for example—that integrate them into a GIS.

The earth can be represented by various models, each of which may provide a different set of coordinates for any given point on the Earth's surface. The simplest model is to assume the earth is a perfect sphere. As more measurements of the earth have accumulated, the models of the earth have become more sophisticated and more accurate. In fact, there are models that apply to different areas of the earth to provide increased accuracy.

Projection is a fundamental component of map making. A projection is a mathematical means of transferring information from a model of the Earth, which represents a three-dimensional curved surface, to a two-dimensional medium—paper or a computer screen. Different projections are used for different types of maps because each projection particularly suits specific uses.

For example, a projection that accurately represents the shapes of the continents will distort their relative sizes. Since much of the information in a GIS comes from existing maps, a GIS uses the processing power of the computer to transform digital information, gathered from sources with different projections and/or different coordinate systems, to a common projection and coordinate system.

SPATIAL ANALYSIS WITH GIS

Given the vast range of spatial analysis techniques that have been developed over the past half century, any summary or review can only cover the subject to a limited depth. This is a rapidly changing field, and GIS packages are increasingly including analytical tools as standard built-in facilities or as optional toolsets, add-ins or 'analysts'.

In many instances such facilities are provided by the original software suppliers, whilst in other cases facilities have been developed and are provided by third parties. Furthermore, many products offer software development kits, programmeming languages and language support, scripting facilities and/or special interfaces for developing one's own analytical tools or variants.

The website Geospatial Analysis and associated book/ebook attempt to provide a reasonably comprehensive guide to the subject. The impact of these myriad paths to perform spatial analysis create a new dimension to business intelligence termed "spatial intelligence" which, when delivered via intranet, democratizes access to operational sorts not usually privy to this type of information.

THREE-DIMENSIONAL GIS

To more realistically analyse the effect of the Earth's terrain, we use three-dimensional models within a GIS. A GIS can display the Earth in realistic, three-dimensional perspective views and animations that convey information more effectively and to wider audiences than traditional, two-dimensional, static maps.

The U.S. Forest Service was offered a land swap by a mining company seeking development rights to a mineral deposit in Arizona's Prescott National Forest. Using a GIS, the USGS and the U.S. Forest Service created perspective views of the area to depict the terrain as it would appear after mining. To assess the potential hazard of landslides both on land and underwater, the USGS generated a three-dimensional image of the San Francisco Bay area. It created the image by mosaicking eight scenes of natural colour composite Landsat 7 enhanced thematic mapper imagery on California fault data using approximately 700 digital elevation models at 1:24,000 scale.

Graphic Display Techniques

Traditional maps are abstractions of the real world; each map is a sampling of important elements portrayed on a sheet of paper with symbols to represent physical objects. People who use maps must interpret these symbols. Topographic maps show the shape of the land surface with contour lines. Graphic display techniques in GISs make relationships among map elements more visible, heightening one's ability to extract and analyse information.

Two types of data were combined in a GIS to produce a perspective view of a part of San Mateo County, Calif. The digital elevation model, consisting of surface elevations recorded on a 30-meter horizontal grid, shows high elevations as white and low elevations as black. The accompanying Landsat thematic mapper image shows a false-colour infrared image of the same area in 30-

meter pixels, or picture elements and combine the two images to produce the three-dimentional image.

Visualization

Maps have traditionally been used to explore the Earth. GIS technology has enhanced the efficiency and analytical power of traditional cartography. As the scientific community recognizes the environmental consequences of human activity, GIS technology is becoming an essential tool in the effort to understand the process of global change. Map and satellite information sources can be combined in models that simulate the interactions of complex natural systems. Through a process known as visualization, a GIS can be used to produce images— not just maps, but drawings, animations, and other cartographic products. These images allow researchers to view their subjects in ways that they never could before. The images often are helpful in conveying the technical concepts of a GIS to nonscientists.

Adding the Element of Time

The condition of the Earth's surface, atmosphere, and subsurface can be examined by feeding satellite data into a GIS. GIS technology gives researchers the ability to examine the variations in Earth processes over days, months, and years. As an example, the changes in vegetation vigor through a growing season can be animated to determine when drought was most extensive in a particular region. The resulting normalized vegetation index represents a rough measure of plant health. Working with two variables over time will allow researchers to detect regional differences in the lag between a decline in rainfall and its effect on vegetation. The satellite sensor used in this analysis is the advanced very high resolution radiometer (AVHRR), which detects the amounts of energy reflected from the Earth's surface at a 1-kilometer resolution twice a day. Other sensors provide spatial resolutions of less than 1 meter.

Serving GIS over the Internet

Through Internet map server technology, spatial data can be accessed and analysed over the Internet. For example, current wildfire perimeters are displayed with a standard web browser, allowing fire managers to better respond to fires while in the field and helping homeowners to take precautionary measures.

The Future of GIS

Environmental studies, geography, geology, planning, business marketing, and other disciplines have benefitted from GIS tools and methods. Together with cartography, remote sensing, global positioning systems, photogrammetry, and geography, the GIS has evolved into a discipline with its own research base known as geographic information sciences. An active GIS market has resulted in lower costs and continual improvements in GIS hardware, software, and data. These developments will lead to a much wider application of the technology throughout government, business, and industry. GIS and related technology will help analyse large datasets, allowing a better understanding of terrestrial processes and human activities to improve economic vitality and environmental quality.

DYNAMICS OF GEOMORPHOLOGY

To place geomorphology upon sound foundations for quantitativeresearch into fundamental principles, it is proposed that geomorphicprocesses be treated as gravitational or molecular shear stressesacting upon elastic, plastic, or fluid earth materials to produce the characteristic varieties of strain, or failure, that constitute weathering, erosion, transportation and deposition.

Shear stresses affecting earth materials are here divided into two major categories: gravitational and molecular. Gravitational stresses activate all downslope movements of matter, hence include all mass movements, all fluvial and glacial processes. Indirect gravitational stresses activate wave-and tide-induced currents and winds. Phenomena of gravitational shear stresses are subdivided according to behavior of rock, soil, ice, water, and air as elastic or plastic solids and viscous fluids. The order of classificationis generally that of decreasing internal resistance to shear and, secondarily, of laminar to turbulent flow.

Molecular stresses are those induced by temperature changes, crystallization and melting, absorption and desiccation, or osmosis. These stresses act in random or unrelated directions with respect to gravity. Surficial creep results from combination of gravitational and molecular stresses on a slope. Chemical processes of solution and acid reaction are considered separately.

A fully dynamic approach requires analysis of geomorphic processes in terms of clearly defined open systems which tend to achieve steady states of operation and are self-regulatory to a largedegree. Formulation of mathematical models, both by rational deduction and empirical analysis of observational data, to relate energy, mass, and time is the ultimate goal of the dynamic approach.

Geomorphology

Geomorphology is the scientific study of landforms and the processes that shape them. Geomorphologists seek to understand why landscapes look the way they do, to understand landform history and dynamics, and to predict future changes through a combination of field observations, physical experiments, and numerical modeling. Geomorphology is practiced within geography, geology, geodesy, engineering geology, archaeology, and geotechnical engineering, and this broad base of interest contributes to a wide variety of research styles and interests within the field.

Overview

The surface of Earth is modified by a combination of surface processes that sculpt landscapes and geologic processes that cause tectonic uplift and subsidence. Surface processes comprise the action of water, wind, ice, fire, and living things on the surface of the Earth, along with chemical reactions that form soils and alter material properties, the stability and rate of change of topography under the force of gravity, and other factors, such as (in the very recent past) human alteration of the landscape.

Many of these factors are strongly mediated by climate. Geologic forcings include the uplift of mountain ranges, the growth of volcanoes, isostatic changes in land surface elevation (sometimes in response to surface processes), and the formation of deep sedimentary basins where the surface of Earth drops and is filled with material eroded from other parts of the landscape.

The Earth surface and its topography therefore are an intersection of climatic, hydrologic, and biologic action with geologic processes.

The broad-scale topographies of Earth illustrate this intersection of surface and subsurface action.

Mountain belts are uplifted due to geologic processes. Denudation of these high uplifted regions produces sediment that is transported and deposited elsewhere within the landscape or off the coast. On progressively smaller scales, similar ideas apply, where individual landforms evolve in response to the balance of additive processes (uplift and deposition) and subtractive processes (subsidence and erosion).

Often, these processes directly affect each other: ice sheets, water, and sediment are all loads that change topography through flexural isostasy. Topography can modify the local climate, for example through orographic precipitation, which in turn modifies the topography by changing the hydrologic regime in which it evolves. Many geomorphologists are particularly interested in the potential for feedbacks between climate and tectonics mediated by geomorphic processes.

In addition to these broad-scale questions, geomorphologists address issues that are more specific and/or more local. Glacial geomorphologists investigate glacial deposits such as moraines, eskers, and proglacial lakes, as well as glacial erosional features, to build chronologies of both small glaciers and large ice sheets and understand their motions and effects upon the landscape.

Fluvial geomorphologists focus on rivers, how they transport sediment, migrate across the landscape, cut into bedrock, respond to environmental and tectonic changes, and interact with humans.

Soils geomorphologists investigate soil profiles and chemistry to learn about the history of a particular landscape and understand how climate, biota, and rock interact. Other geomorphologists study how hillslopes form and change. Still others investigate the relationships between ecology and geomorphology. Because geomorphology is defined to comprise everything related to the surface of Earth and its modification, it is a broad field with many facets.

Practical applications of geomorphology include hazard assessment (such as landslide prediction and mitigation), river control and stream restoration, and coastal protection.

History

With some notable exceptions, geomorphology is a relatively young science, growing along with interest in other aspects of the Earth Sciences in the mid 19th century. This section provides a very brief outline of some of the major figures and events in its development.

Ancient Geomorphology

Perhaps the earliest one to devise a theory of geomorphology was the polymath Chinese scientist and statesman Shen Kuo (1031-1095 AD). This was based on his observation of marine fossil shells in a geological stratum of a mountain hundreds of miles from the Pacific Ocean. Noticing bivalve shells running in a horizontal span along the cut section of a cliffside, he theorized that the cliff was once the pre-historic location of a seashore that had shifted hundreds of miles over the centuries. He inferred that the land was reshaped and formed by soil erosion of the mountains and by deposition of silt, after observing strange natural erosions of the Taihang Mountains and the Yandang Mountain near Wenzhou. Furthermore, he promoted the theory of gradual climate change over centuries of time once ancient petrified bamboos were found to be preserved underground in the dry, northern climate zone of *Yanzhou*, which is now modern day Yan'an, Shaanxi province.

Early Modern Geomorphology

The first use of the word geomorphology was likely to be in the German language when it appeared in Laumann's 1858 work. Keith Tinkler has suggested that the word came into general use in English, German and French after John Wesley Powell and W. J. McGee used it in the International Geological Conference of 1891.

An early popular geomorphic model was the *geographical cycle* or the *cycle of erosion,* developed by William Morris Davis between 1884 and 1899. The cycle was inspired by theories of uniformitarianism first formulated by James Hutton (1726–1797). Concerning valley forms, uniformitarianism depicted the cycle as a sequence in which a river cuts a valley more and more deeply, but then erosion of side valleys eventually flatten the terrain again, to a lower elevation.

Tectonic uplift could start the cycle over. Many studies in geomorphology in the decades following Davis' development of his theories sought to fit their ideas into this framework for broad scale landscape evolution, and are often today termed "Davisian".

Davis' ideas have largely been superseded today, mainly due to their lack of predictive power and qualitative nature, but he remains an extremely important figure in the history of the subject.

In the 1920s, Walther Penck developed an alternative model to Davis', believing that landform evolution was better described as a balance between ongoing processes of uplift and denudation, rather than Davis' single uplift followed by decay. However, due to his relatively young death, disputes with Davis and a lack of English translation of his work his ideas were not widely recognised for many years.

These authors were both attempting to place the study of the evolution of the Earth's surface on a more generalized, globally relevant footing than had existed before. In the earlier parts of the 19th century, authors-especially in Europe-had tended to attribute the form of landscape to local climate, and in particular to the specific effects of glaciation and periglacial processes. In contrast, both Davis and Penck were seeking to emphasize the importance of evolution of landscapes through time and the generality of Earth surface processes across different landscapes under different conditions.

Quantitative Geomorphology

While Penck and Davis and their followers were writing and studying primarily in Western Europe, another, largely separate, school of geomorphology was developed in the United States in the middle years of the 20th century.

Following the early trailblazing work of Grove Karl Gilbert around the turn of the 20th century, a group of natural scientists, geologists and hydraulic engineers including Ralph Alger Bagnold, John Hack, Luna Leopold, Thomas

Maddock and Arthur Strahler began to research the form of landscape elements such as rivers and hillslopes by taking systematic, direct, quantitative measurements of aspects of them and investigating the scaling of these measurements.

These methods began to allow prediction of the past and future behavior of landscapes from present observations, and were later to develop into what the modern trend of a highly quantitative approach to geomorphic problems. Quantitative geomorphology can involve fluid dynamics and solid mechanics, geomorphometry, laboratory studies, field measurements, theoretical work, and full landscape evolution modeling. These approaches are used to understand weathering and the formation of soils, sediment transport, landscape change, and the interactions between climate, tectonics, erosion, and deposition.

6

Remote Sensing System

The term 'remote sensing' can generally be defined as "the science and art of obtaining information about an object, area, or phenomenon through the analysis of data acquired by a device not in contact with the object, area, or phenomenon under investigation." This broad definition includes cameras, ocean buoys, and RADAR devices. These general remote sensing devices have undergone adaptations over the last century to make observations of the Earth and its physical process from aircraft and satellites.

Airplanes have been used since the 1930s to carry cameras and sensors to study the earth. Cameras collect images of part of Earth's surface, with the final aerial photograph usually consisting of a series of overlapping vertical photos that form the basis for mapping. Airplanes are also used to carry sensors.

For example, the Side-Looking Airborne Radar (SLAR) instrument is used by the United States Geological Survey (USGS) for various projects in the conterminous United States and Alaska to map geologic features, detect mineral and energy reserves, and identify potential environmental hazards.

Satellites are also used to collect images and data about the earth. Earth-observing satellites, as they are referred to, carry sensors, which are capable of recording wavelengths across the entire electromagnetic spectrum, from infrared to visible radiation. Some satellites carry sensors that collect data passively by recording radiation that is radiated or reflected from Earth's surface or atmosphere. Other satellites collect data actively by emitting radiation and then recording what is reflected back to them from Earth's surface or atmosphere.

APPLICABLE TO REMOTE SENSING

The Daubert standards reviewed above will only be applied to remotely-sensed data presented through expert testimony. Remote sensing evidence will be subject to several FRE, which are applicable whether or not an expert is called to testify.

Relevancy, Authentication, and Foundation Any evidence, scientific or otherwise, must be found relevant to the case, meaning that it must make a consequential fact more or less probable than would be deemed otherwise. If used to aid witness testimony, the map must help the trier of fact understand the testimony. Once evidence is found to be relevant, it must be authenticated. Extrinsic authentication is necessary unless the map fulfills one of the self-authentication exceptions listed in Rule 902 of the FRE. A map published by the government, for instance, is self-authenticating under Rule 902(5).

Finally, the evidence must have an adequate foundation; it must be accurate and reliable. If accuracy cannot be confirmed, courts will not admit the evidence. Of these provisions, the main evidentiary hurdle for digital maps is reliability. Courts will ask where the information in the map originated, how the information was transformed into digital form, and how the map itself was created. Since computers create digital maps, the maps will face reliability challenges as computer evidence. Courts, for instance, will inquire into "computer programming errors, equipment malfunction, data entry errors, and the volume of electronic data." Courts will also closely consider the authenticity of digital maps, particularly where the map does not meet one of the aforementioned self-authentication exceptions. As such, courts will follow Rule 901(a), requiring proof that the evidence is what its proponent claims it to be. According to Rule 901(b)(9), parties must prove that evidence encompassing a process or system, such as maps depicting remotely-sensed data, must produce an accurate result. To satisfy these rules, the experts who collected the remotely-sensed data should describe how the process operates and their involvement. Experts should also reference the data to ground information ('ground-truthing'), aerial photographs, and other maps. Logs and records of the progression from collection to presentation of the data would also verify authenticity. Technologies including steganography and cyclic redundant checksum are continually being developed to assist in ensuring the authenticity of digital imagery.

Hearsay Issues

If a map, chart, or other media is admitted to make an assertion, the evidence may be objected to on hearsay grounds. For example, remotely-sensed data could be used to create a map depicting high levels of pollution in a stream adjacent to the defendant's property. If the map is admitted to assert that the defendant caused such pollution, it may meet with a hearsay objection. If the evidence is found to be hearsay, it will only be admissible if it can be categorized as an exception to the hearsay rules. For example, Rules 803(6) and 803(8) will allow the admission of hearsay evidence that was generated by computer for use as a business or public record.

Data Characterization

A final set of rules that may pertain to the use of remotely-sensed data involve the presentation of the evidence in the courtroom. Rule 1006 allows the admission of charts, summaries, and calculations that depict a body of data too voluminous to itself be admitted into evidence for practical reasons. To avoid potential problems with admission under this rule, experts should testify that the data was correctly translated into these summary forms. If the evidence is admitted without the verification of expert testimony, Rule 1002 requires that the underlying data be admissible. For example, if a chart includes data derived from satellite photos, courts or opposing attorneys could bar the admission of the chart if the original photos do not also meet the standards of admissibility.

Constitutional Hurdles

Besides Daubert and the FRE, the Constitution presents another obstacle that remote sensing data must overcome for admission into federal courts. The main constitutional issues facing remote sensing data are allegations of invasions of privacy and warrantless searches. The Fourth Amendment states that "the right of the people to be secure in their persons, houses, papers, and effects, against unreasonable searches and seizures, shall not be violated." Two Supreme Court cases, Dow Chemical Co. v. United States and Kyllo v. United States, address the application of the Constitution to remote sensing data.

In Dow Chemical Co., the Court held that enhanced aerial photographs of an industrial facility taken by the EPA were admissible under the Fourth Amendment. The Court found that though commercial areas receive constitutional privacy protection, this protection does not extend to the outdoor areas of industrial complexes.

The Court also found that homes and their outside areas receive a higher level of protection than commercial areas. Still, in dicta the Court stated, "surveillance of private property by using highly sophisticated surveillance equipment not generally available to the public, such as satellite technology, might be constitutionally proscribed absent a warrant." The Court feared that technology providing information not available to the naked eye would reveal intimate details, for example, imaging that could reveal actions occurring inside a building (e.g., conversations behind closed doors or people transporting documents). Despite this concern, the Court noted that photos enhancing human vision were still admissible, provided that they do not reveal such intimate details.

The Supreme Court's latest decision regarding remote sensing data's privacy and search issues is Kyllo v. United States. Kyllo involved a police officer who used a thermal imaging device to detect heat emissions from a suspect's home. Declaring this search unconstitutional, the Court held that when "the Government uses a device that is not in general public use, to explore details of the home that would previously have been unknowable without physical intrusion, the surveillance" is unconstitutional. As in Dow Chemical Co., the Court emphasized that homes receive a high level of privacy protection under the Constitution. The Court held that, in the home, "all details are intimate details," strongly indicating that any information obtained by remote sensing data from a home's interior without a warrant would be inadmissible.

The Court did not define "general use" technology in either Dow Chemical Co. or Kyllo. Lower courts are left to speculate on what level of use might rise to this standard. For example, remote sensors that track wetland deterioration might be deemed "general use" technology if they are routinely used by the government, or if the public accepted their use. But if the device determined that someone illegally filled in a wetland in his or her backyard, that information could be inadmissible. The main lesson that can clearly be drawn from Dow Chemical Co. and Kyllo is that, in the absence of a warrant, remote sensing data will only gain courtroom admission if it does not include intimate details of commercial activity or any details from private homes.

AN OVERVIEW OF REMOTE SENSING

Remote sensing is the examination or the gathering of information about a place from a distance. Such examination can occur with devices (*e.g.*-cameras) based on the ground, and/or sensors or cameras based on ships, aircraft, satellites, or other spacecraft. Today, the data obtained is usually stored and manipulated using computers. The most common software used in remote sensing is ERDAS Imagine, ESRI, MapInfo, and ERMapper.

Modern remote sensing began in 1858 when Gaspard-Felix Tournachon first took aerial photographs of Paris from a hot air balloon. Remote sensing continued to grow from there; one of the first planned uses of remote sensing occurred during the U.S. Civil War when messenger pigeons, kites, and unmanned balloons were flown over enemy territory with cameras attached to them. The first governmental-organized air photography missions were developed for military surveillance during World Wars I and II but reached a climax during the Cold War.

Today, small remote sensors or cameras are used by law enforcement and the military in both manned and unmanned platforms to gain information about an area. Today's remote sensing imaging also includes infra-red, conventional air photos, and Doppler radar. In addition to these tools, satellites were developed during the late 20th century and are still used today to gain information on a global scale and even information about other planets in the solar system. For example, the Magellan probe is a satellite that has used remote sensing technologies to create topographic maps of Venus.

TYPES OF REMOTE SENSING DATA

The types of remote sensing data vary but each plays a significant role in the ability to analyze an area from some distance away. The first way to gather remote sensing data is through radar. Its most important uses are for air traffic control and the detection of storms or other potential disasters. In addition, Doppler radar is a common type of radar used in detecting meteorological data but is also used by law enforcement to monitor traffic and driving speeds. Other types of radar are also used to create digital models of elevation.

Another type of remote sensing data comes from lasers. These are often used in conjunction with radar altimeters on satellites to measure things like wind speeds and their direction and the direction of ocean currents. These altimeters are also useful in seafloor mapping in that they are capable of measuring bulges of water caused by gravity and the varied seafloor topography. These varied ocean heights can then be measured and analyzed to create seafloor maps. Also common in remote sensing is LIDAR-Light Detection and Ranging. This is most famously used for weapons ranging but can also be used to measure chemicals in the atmosphere and heights of objects on the ground.

Other types of remote sensing data include stereographic pairs created from multiple air photos (often used to view features in 3-D and/or make topographic maps), radiometers and photometers which collect emitted radiation common in infra-red photos, and air photo data obtained by earth-viewing satellites such as those found in the Landsat programme.

Applications of Remote Sensing

As with its varied types of data, the specific applications of remote sensing are diverse as well. However, remote sensing is mainly conducted for image processing and interpretation. Image processing allows things like air photos and satellite images to be manipulated so they fit various project uses and/or to create maps. By using image interpretation in remote sensing an area can be studied without being physically present there. The processing and interpretation of remote sensing images also has specific uses within various fields of study. In geology, for instance, remote sensing can be applied to analyze and map large, remote areas. Remote sensing interpretation also makes it easy for geologists in this case to identify an area's rock types, geomorphology, and changes from natural events such as a flood or landslide. Remote sensing is also helpful in studying vegetation types. Interpretation of remote sensing images allows physical and biogeographers, ecologists, those studying agriculture, and foresters to easily detect what vegetation is present in certain areas, its growth potential, and sometimes what conditions are conducive to its being there.

Additionally, those studying urban and other land use applications are also concerned with remote sensing because it allows them to easily pick out which land uses are present in an area. This can then be used as data in city planning applications and the study of species habitat, for example. Finally, remote sensing plays a significant role in GIS.

Its images are used as the input data for the raster-based digital elevation models (abbreviated as DEMs)-a common type of data used in GIS. The air photos taken during remote sensing applications are also used during GIS digitizing to create polygons, which are later put into shapefiles to create maps. Because of its varied applications and ability to allow users to collect, interpret, and manipulate data over large often not easily accessible and sometimes dangerous areas, remote sensing has become a useful tool for all geographers, regardless of their concentration.

PROCESS OF GATHERING INFORMATION IN REMOTE SENSING

Remote sensing refers to the process of gathering information about an object, at a distance, without touching the object itself. The most common remote sensing method that comes to most people's minds is the photographic image of an object taken with a camera. Remote sensing has evolved into much more than looking at objects with our eyes. It now includes using instruments, which can measure attributes about objects which unaided human eyes can't see or sense.

Some other definitions of Remote Sensing are: "Photogrammetry and Remote Sensing are the art, science and technology of obtaining reliable information about physical objects and the environment, through a process of recording, measuring and interpreting imagery and digital representations of energy patterns derived from noncontact sensor systems".

"Remote sensing may be broadly defined as the collection of information about an object without being in physical contact with the object. Aircraft and satellites are the common platforms from which remote sensing observations are made. The term remote sensing is restricted to methods that employ electromagnetic energy as the means of detecting and measuring target characteristics".

"Remote sensing is the art and science of obtaining information from a distance, *i.e.* obtaining information about objects or phenomena without being in physical contact with them. The science of remote sensing provides the instruments and theory to understand how objects and phenomena can be detected. The art of remote sensing is in the development and use analysis techniques to generate useful information".

History

In 1858 a French photographer, Gaspaed Felix Tournachon was the first to take aerial photos from a tethered balloon. A few years later in 1861, aerial photographs became a tool for military intelligence during the civil war. Aerial photographs were also taken from cameras mounted in kites (1858), and on carrier pigeons (1903). In 1909 Wilber Wright flew the first airplane to take the first photographs in flight. The first aerial photographs used in the process of creating maps was presented in a paper in 1913, by Captain Tardivo at a meeting of the International Society for Photogrammetry. Military aerial photos were used on a large scale during World War I. The military trained hundreds of people to process and interpret aerial reconnaissance photos. The French aerial units developed 56,000 photos in four days during the Meuse-Argonne offensive in 1918. After World War I and through the 1930's, commercial aerial survey companies employed many former military personnel to process aerial photos to produce maps such as topographic maps, forest management maps, and soil maps.

World War II saw the development of colour-infrared film for the US Army in 1942. These images were used to detect enemy forces and equipment that were camouflaged. A majority of Allied intelligence gathered about the enemy during this war was the direct result of aerial photoreconnaissance.

The U.S. military and other government agencies such as National Aeronautics and Space Administration (NASA) continued to develop the use of remote sensing during the cold war years. The 1960's also saw the expansion and development of earth remote sensing from space. The first military space photo reconnaissance satellite, Corona, was launched in 1960. Corona took pictures of the Soviet Union and its allies using photographic film. The exposed film was then transferred into unmanned recovery vehicles in space. The recovery vehicles then de-orbited and returned to earth by parachute carrying the film, which was then processed and analyzed in the lab. The first series of weather satellites called the Television Infrared Observation Satellites (TIROS) began launching in 1960. NASA continued collecting images for its earth observation surveys, from outer space, with the Apollo and Gemini spacecraft.

Scores of U.S. meteorological and earth observation satellites were launched during the 1970's. Also during the 1970's manned spacecraft such as the Skylab space station collected images of earth from outer space. In 1972 Landsat-1, with an original resolution of only 80 meters was the first satellite launched into space for nonmilitary earth resource observation. Landsat contained sensors capable of taking multispectral digital images. U.S. military photoreconnaissance satellites have been keep secrete and unavailable to the general public. Starting in 1976 the U.S. military started deploying more sophisticated high-resolution satellites capable of relaying digital images to earth. Eight Keyhole-11 satellites were launched between 1976 and 1988. Three improved Keyhole-11B satellites were launched between 1992 and 1996. They are able to produce images with estimated resolutions of nearly ten centimeters (four inches).

Nonmilitary satellite images have been used to monitor the degradation and pollution of the environment. These images also can be used to assess the damage of floods and natural disasters, assist in forecasting the weather, locate minerals and oil reserves, locate fish stocks, monitor ocean currents, assist in land use mapping and planning, produce geologic maps, and monitor range, forestry and agricultural resources.

PRINCIPLES OF REMOTE SENSING

We perceive the surrounding world through our five senses. Some senses (touch and taste) require contact of our sensing organs with the objects. However, we acquire much information about our surrounding through the senses of sight and hearing which do not require close contact between the sensing organs and the external objects. In another word, we are performing Remote Sensing all the time.

Generally, Remote sensing refers to the activities of recording/observing/ perceiving (sensing) objects or events at far away (remote) places. In remote sensing, the sensors are not in direct contact with the objects or events being observed. The information needs a physical carrier to travel from the objects/ events to the sensors through an intervening medium. The electromagnetic radiation is normally used as an information carrier in remote sensing. The output of a remote sensing system is usually an image representing the scene being observed. A further step of image analysis and interpretation is required in order to extract useful information from the image. The human visual system is an example of a remote sensing system in this general sense.

In a more restricted sense, remote sensing usually refers to the technology of acquiring information about the earth's surface (land and ocean) and atmosphere using sensors onboard airborne (aircraft, balloons) or spaceborne (satellites, space shuttles) platforms.

Satellite Remote Sensing

In this CD, you will see many remote sensing images around Asia acquired by earth observation satellites. These remote sensing satellites are equipped with sensors looking down to the earth. They are the "eyes in the sky" constantly observing the earth as they go round in predictable orbits.

Effects of Atmosphere

In satellite remote sensing of the earth, the sensors are looking through a layer of atmosphere separating the sensors from the Earth's surface being observed. Hence, it is essential to understand the effects of atmosphere on the electromagnetic radiation travelling from the Earth to the sensor through the atmosphere. The atmospheric constituents cause wavelength dependent absorption and scattering of radiation. These effects degrade the quality of images. Some of the atmospheric effects can be corrected before the images are subjected to further analysis and interpretation.

A consequence of atmospheric absorption is that certain wavelength bands in the electromagnetic spectrum are strongly absorbed and effectively blocked by the atmosphere. The wavelength regions in the electromagnetic spectrum usable for remote sensing are determined by their ability to penetrate atmosphere. These regions are known as the atmospheric transmission windows. Remote sensing systems are often designed to operate within one or more of the atmospheric windows. These windows exist in the microwave region, some wavelength bands in the infrared, the entire visible region and part of the near ultraviolet regions. Although the atmosphere is practically transparent to x-rays and gamma rays, these radiations are not normally used in remote sensing of the earth.

Optical and Infrared Remote Sensing

In Optical Remote Sensing, optical sensors detect solar radiation reflected or scattered from the earth, forming images resembling photographs taken by a camera high up in space. The wavelength region usually extends from the visible and near infrared (commonly abbreviated as VNIR) to the short-wave infrared (SWIR). Different materials such as water, soil, vegetation, buildings and roads reflect visible and infrared light in different ways. They have different colours and brightness when seen under the sun. The interpretation of optical images require the knowledge of the spectral reflectance signatures of the various materials (natural or man-made) covering the surface of the earth.

There are also infrared sensors measuring the thermal infrared radiation emitted from the earth, from which the land or sea surface temperature can be derived.

Microwave Remote Sensing

There are some remote sensing satellites which carry passive or active microwave sensors. The active sensors emit pulses of microwave radiation to illuminate the areas to be imaged. Images of the earth surface are formed by measuring the microwave energy scattered by the ground or sea back to the sensors. These satellites carry their own "flashlight" emitting microwaves to illuminate their targets. The images can thus be acquired day and night. Microwaves have an additional advantage as they can penetrate clouds. Images can be acquired even when there are clouds covering the earth surface.

A microwave imaging system which can produce high resolution image of the Earth is the synthetic aperture radar (SAR). The intensity in a SAR image depends on the amount of microwave backscattered by the target and received by the SAR antenna. Since the physical mechanisms responsible for this backscatter is different for microwave, compared to visible/infrared radiation, the interpretation of SAR images requires the knowledge of how microwaves interact with the targets.

Remote Sensing Images

Remote sensing images are normally in the form of digital images. In order to extract useful information from the images, image processing techniques may be employed to enhance the image to help visual interpretation, and to correct or restore the image if the image has been subjected to geometric distortion, blurring or degradation by other factors. There are many image analysis techniques available and the methods used depend on the requirements of the specific problem concerned. In many cases, image segmentation and classification algorithms are used to delineate different areas in an image into thematic classes. The resulting product is a thematic map of the study area. This thematic map can be combined with other databases of the test area for further analysis and utilization.

Visual System

Passive Remote Sensing: The eyes passively senses the radiation reflected or emitted from the object. The sensing system depends on an external source of illumination.

The human visual system is an example of a remote sensing system in the general sense. The sensors in this example are the two types of photosensitive cells, known as the cones and the rods, at the retina of the eyes. The cones are responsible for colour vision. There are three types of cones, each being sensitive to one of the red, green, and blue regions of the visible spectrum. Thus, it is not coincidental that the modern computer display monitors make use of the same three primary colours to generate a multitude of colours for displaying colour images. The cones are insensitive under low light illumination condition, when their jobs are taken over by the rods. The rods are sensitive only to the total light intensity. Hence, everything appears in shades of grey when there is insufficient light. As the objects/events being observed are located far away from the eyes, the information needs a carrier to travel from the object to the eyes. In this case, the information carrier is the visible light, a part of the electromagnetic spectrum. The objects reflect/scatter the ambient light falling onto them.

Part of the scattered light is intercepted by the eyes, forming an image on the retina after passing through the optical system of the eyes. The signals generated at the retina are carried via the nerve fibres to the brain, the central processing unit (CPU) of the visual system. These signals are processed and interpreted at the brain, with the aid of previous experiences. When operating in this mode, the visual system is an example of a "Passive Remote Sensing" system which depends on an external source of energy to operate. We all know that this system won't work in darkness. However, we can still see at night if we provide our own source of illumination by carrying a flashlight and shining the beam towards the object we want to observe. In this case, we are performing "Active Remote Sensing", by supplying our own source of energy for illuminating the objects.

The Planet Earth

The planet Earth is the third planet in the solar system located at a mean distance of about 1.50 x 10^8 km from the sun, with a mass of 5.97 x 10^{24} kg. Descriptions of the shape of the earth have evolved from the flat-earth model, spherical model to the currently accepted ellipsoidal model derived from accurate ground surveying and satellite measurements. A number of reference ellipsoids have been defined for use in identifying the three dimensional coordinates (*i.e.* position in space) of a point on or above the earth surface for the purpose of surveying, mapping and navigation. The reference ellipsoid in the World Geodetic System 1984 (WGS-84) commonly used in satellite Global Positioning System (GPS) has the following parameters:

- Equatorial Radius = 6378.1370 km
- Polar Radius = 6356.7523 km

The earth's crust is the outermost layer of the earth's land surface. About 29.1% of the earth's crust area is above sea level. The rest is covered by water. A layer of gaseous atmosphere envelopes the earth's surface.

The Earth's Atmosphere

The earth's surface is covered by a layer of atmosphere consisting of a mixture of gases and other solid and liquid particles. The gaseous materials extend to several hundred kilometers in altitude, though there is no well defined boundary for the upper limit of the atmosphere. The first 80 km of the atmosphere contains more than 99% of the total mass of the earth's atmosphere.

Vertical Structure of the Atmosphere

The vertical profile of the atmosphere is divided into four layers: troposphere, stratosphere, mesosphere and thermosphere. The tops of these layers are known as the tropopause, stratopause, mesopause and thermopause, respectively.

- Troposphere: This layer is characterized by a decrease in temperature with respect to height, at a rate of about 6.5ºC per kilometer, up to a height of about 10 km. All the weather activities (water vapour, clouds, precipitation) are confined to this layer. A layer of aerosol particles normally exists near to the earth surface. The aerosol concentration decreases nearly exponentially with height, with a characteristic height of about 2 km.
- Stratosphere: The temperature at the lower 20 km of the stratosphere is approximately constant, after which the temperature increases with height, up to an altitude of about 50 km. Ozone exists mainly at the stratopause. The troposphere and the stratosphere together account for more than 99% of the total mass of the atmosphere.
- Mesosphere: The temperature decreases in this layer from an altitude of about 50 km to 85 km.
- Thermosphere: This layer extends from about 85 km upward to several hundred kilometers. The temperature may range from 500 K to 2000 K. The gases exist mainly in the form of thin plasma, *i.e.* they are ionized due to bombardment by solar ultraviolet radiation and energetic cosmic rays.

The term upper atmosphere usually refers to the region of the atmosphere above the troposphere.

Many remote sensing satellites follow the near polar sun-synchronous orbits at a height around 800 km, which is well above the thermopause.

Atmospheric Constituents

The atmosphere consists of the following components:

- Permanent Gases: They are gases present in nearly constant concentration, with little spatial variation. About 78% by volume of the atmosphere is nitrogen while the life-sustaining oxygen occupies 21%. The remaining one percent consists of the inert gases, carbon dioxide and other gases.
- Gases with Variable Concentration: The concentration of these gases may vary greatly over space and time. They consist of water vapour, ozone, nitrogeneous and sulphurous compounds.
- Solid and liquid particulates: Other than the gases, the atmosphere also

contains solid and liquid particles such as aerosols, water droplets and ice crystals. These particles may congregate to form clouds and haze.

Electromagnetic Waves

Electromagnetic waves are energy transported through space in the form of periodic disturbances of electric and magnetic fields. All electromagnetic waves travel through space at the same speed, c = 2.99792458 x 10^8 m/s, commonly known as the speed of light. An electromagnetic wave is characterized by a frequency and a wavelength. These two quantities are related to the speed of light by the equation:

Speed of light = frequency x wavelength

The frequency (and hence, the wavelength) of an electromagnetic wave depends on its source. There is a wide range of frequency encountered in our physical world, ranging from the low frequency of the electric waves generated by the power transmission lines to the very high frequency of the gamma rays originating from the atomic nuclei. This wide frequency range of electromagnetic waves constitute the Electromagnetic Spectrum.

The Electromagnetic Spectrum

The electromagnetic spectrum can be divided into several wavelength (frequency) regions, among which only a narrow band from about 400 to 700 nm is visible to the human eyes. Note that there is no sharp boundary between these regions. The boundaries shown in the above figures are approximate and there are overlaps between two adjacent regions.

Wavelength units: 1 mm = 1000 μm; 1 μm = 1000 nm.

- Radio Waves: 10 cm to 10 km wavelength.
- Microwaves: 1 mm to 1 m wavelength. The microwaves are further divided into different frequency (wavelength) bands: (1 GHz = 10^9 Hz)
  - P band: 0.3-1 GHz (30-100 cm)
  - L band: 1-2 GHz (15-30 cm)
  - S band: 2-4 GHz (7.5-15 cm)
  - C band: 4-8 GHz (3.8-7.5 cm)
  - X band: 8-12.5 GHz (2.4-3.8 cm)
  - Ku band: 12.5-18 GHz (1.7-2.4 cm)
  - K band: 18-26.5 GHz (1.1-1.7 cm)
  - Ka band: 26.5-40 GHz (0.75-1.1 cm)
- Infrared: 0.7 to 300 μm wavelength. This region is further divided into the following bands:
  - Near Infrared (NIR): 0.7 to 1.5 μm.
  - Short Wavelength Infrared (SWIR): 1.5 to 3 μm.

- Mid Wavelength Infrared (MWIR): 3 to 8 μm.
- Long Wanelength Infrared (LWIR): 8 to 15 μm.
- Far Infrared (FIR): longer than 15 μm.

The NIR and SWIR are also known as the Reflected Infrared, referring to the main infrared component of the solar radiation reflected from the earth's surface. The MWIR and LWIR are the Thermal Infrared.

- Visible Light: This narrow band of electromagnetic radiation extends from about 400 nm (violet) to about 700 nm (red). The various colour components of the visible spectrum fall roughly within the following wavelength regions:
  - Red: 610-700 nm
  - Orange: 590-610 nm
  - Yellow: 570-590 nm
  - Green: 500-570 nm
  - Blue: 450-500 nm
  - Indigo: 430-450 nm
  - Violet: 400-430 nm
- Ultraviolet: 3 to 400 nm
- X-Rays and Gamma Rays

SIGNATURE IN REMOTE SENSING

The knowledge of spectral signatures is essential for exploiting the potential of remote sensing techniques. This knowledge enables one to identify and classify the objects of agricultural resources. It is also required for interpretation of all remotely sensed data, especially in agricultural resource data whether the interpretation is carried out visually or using digital techniques.

It also helps us in specifying requirements for any remote sensing mission *e.g.* which optimal wave length bands to be used or which type of sensor will be best suited for a particular task (agricultural survey). All objects of agricultural resource on the surface of the earth have characteristic spectral signatures. The average spectral reflectance curves (or) spectral signatures for three typical earth's features; vegetation, soil and water. The spectral reflectance curves for vigorous vegetation manifests the "Peak-and valley" configuration. The valleys in the visible portion of the spectrum are indicative of pigments in plant leaves. Dips in reflectance that can be sent at wavelengths of 0.65 mm, 1.4 mm and 1.9 mm are attributable to absorption of water by leaves. The soil curves show a more regular variation of reflectance.

Factors that evidently affect soil reflectance are moisture content, soil texture, surface roughness and presence of organic matter. The water curves shows that from about 0.5 mm, reduction in reflectance with increasing wavelength, so that in the near infrared range, the reflectance of deep clear water is virtually zero (Mather, 1987) However, the spectral reflectance of water is significantly affected by the presence of dissolved and suspended organic and inorganic material and by the depth of the water body. Determinations of spectral signatures implies basic under standing of interaction of electromagnetic radiation with agricultural resources objects. This is also necessary for analysing and designing sensor systems for agricultural survey.

Sensor Systems in Remote Sensing

In remote sensing the acquisition of data is depending upon the sensor system used. Various remote sensing platforms (Aircraft, Satellite) are equipped with different sensor systems. Sensor is a device that receives electromagnetic radiation, converts it into a signal and presents it in a form suitable of obtaining information about the land or earth resource as used by an information gathering system. Sensor can be grouped, either on the basic of energy source. They are as classified.

Active sensor: An active sensor operates by emitting its own energy, which is needed to detect the various phenomena (*e.g.* RADAR, camera with a flash gun).

Passive sensor: The operation of passive sensor is dependent of the existing sources of energy, like sun (*e.g.* photographic systems, multispectral scanners).

The given sensor system of camera are in agricultural survey.

Photographic Cameras

The photographic system, having conventional camera with black and white photography, is the oldest and probably, so far, the most widely used sensor for recording information about ground object. Photographic cameras have been successfully used in aircraft platform remote sensing. In this system, the information is limited to size and shape, as the films used are sensitive only to visible region of spectrum. The response of black & white films is about 0.4-0.7 mm for infrared imagery, films with response extending up to 0.9 mm are available.

Return Beam Vidicon (RBM)

This is very similar to a television camera. In such a system, a fixed camera lens on a photosensitive semi-transparent sheet forms the ground image. This image is created on the surface as electrical change or potential. The TV cameras are the best example of high resolution, operated in space for resource survey was the RBV used in LAND SAT series. On LAND SAT I, II and III RBV cameras were used, each corresponding to a different wavelength band 0.475-0.585 mm (green), 0.580-0.690 mm (red) and 0.690-0.830 mm (near infrared). The Indian experimental remote sensing satellite, Bhaskara-I and II carried a two-band TV camera system, Multispectral imagery was produced in LAND SAT and Bhaskara by using separate camera tubes of each band and selecting the spectral band with appropriate filters.

Optical-mechanical Scanners

This imaging system has the advantage that any set of desired spectral bands can be selected with appropriate filter and detector combinations. The mostly widely used sensor in this category is the MSS on LAND SAT series. MSS has four spectral bands, covering form 0.5-to 1.1 mm region. MSS operates on the principle of scanning successive lines at right angles to the flight path by means of a rotation or oscillating optical system. The radiation levels along the lines are recorded by appropriate sensor elements. When used in the visible band, the collected light can be split by the optics and separately filtered and recorded, giving simultaneous multispectral recording from the one instrument. MSS can record in any part of ultraviolet to near IR window. They are use also in the thermal IR windows.

Radar and Microwave Sensors

The acquisition of data in microwave region has been possible since 1950s but its application to natural resources is considerably less developed, as compared to the visible and IR image interpretations. Microwave sensors have distinct advantages because they are unaffected by atmospheric conditions and are thus able to penetrated smoke, clouds, haze and snow. Under this system, Plan Position Indicator (PPI), Side Looking Air borne Radar (SLAR) and Synthetic Aperture Radar (SAR) can be grouped. These systems offer day and night as well as all weather capability and ability to penetrate a cover of vegetation.

Advance Remote Sensors

Linear Imaging and Self Scanning Sensors (LISS) are the advanced imaging systems. This type of scanning sensor are used an array of solid-state devices. The array may be made of photo-diodes, phototransistors or Charge-Coupled Devices (CCDs). In the LISS, the optics focuses a strip of terrain in the cross-track into the sensor array. The image from each detector is stored and shifted out sequentially to receive a video signal. The SPOT (Satellite Probatorie d' Observation de la Terra) and IRS (Indian Remote Sensing Satellite) series carry such solid-state sensor systems, which are also known as push-broom scanners. The IRS IC most advanced satellite, carries an improved sensor system. Besides carrying a sophisticated LISS-III camera, it has a Panchromatic camera (PAN) and a Wide Field Sensor (WiFS). The PAN has been designed to provide data with a spatial resolution of 5.8m in stereo mode, with a ground swath of 70km, whereas WiFS provides data in two spectral bands, with a spectral resolution of 188m and a ground swath of 180km.

DATA ACQUISITION TECHNIQUES IN REMOTE SENSING

Remote sensing is the acquisition of information about an object or phenomenon, without making physical contact with the object. In modern usage, the term generally refers to the use of aerial sensor technologies to detect and classify objects on Earth (both on the surface, and in the atmosphere and oceans) by means of propagated signals (*e.g.* electromagnetic radiation emitted from aircraft or satellites).

There are two main types of remote sensing: passive remote sensing and active remote sensing. Passive sensors detect natural radiation that is emitted or reflected by the object or surrounding area being observed. Reflected sunlight is the most common source of radiation measured by passive sensors. Examples of passive remote sensors include film photography, infrared, charge-coupled devices, and radiometers.

Active collection, on the other hand, emits energy in order to scan objects and areas whereupon a sensor then detects and measures the radiation that is reflected or backscattered from the target.

RADAR is an example of active remote sensing where the time delay between emission and return is measured, establishing the location, height, speed and direction of an object.

Remote sensing makes it possible to collect data on dangerous or inaccessible areas.

Remote sensing applications include monitoring deforestation in areas such as the Amazon Basin, glacial features in Arctic and Antarctic regions, and depth sounding of coastal and ocean depths.

Military collection during the cold war made use of stand-off collection of data about dangerous border areas. Remote sensing also replaces costly and slow data collection on the ground, ensuring in the process that areas or objects are not disturbed.

Orbital platforms collect and transmit data from different parts of the electromagnetic spectrum, which in conjunction with larger scale aerial or ground-based sensing and analysis, provides researchers with enough information to monitor trends such as El Niño and other natural long and short term phenomena. Other uses include different areas of the earth sciences such as natural resource management, agricultural fields such as land usage and conservation, and national security and overhead, ground-based and stand-off collection on border areas. By satellite, aircraft, spacecraft, buoy, ship, and helicopter images, data is created to analyze and compare things like vegetation rates, erosion, pollution, forestry, weather, and land use. These things can be mapped, imaged, tracked and observed. The process of remote sensing is also helpful for city planning, archaeological investigations, military observation and geomorphological surveying.

Data Acquisition Techniques

The basis for multispectral collection and analysis is that of examined areas or objects that reflect or emit radiation that stand out from surrounding areas.

PASSIVE REMOTE SENSING

Passive systems collect data from energy that is reflected or radiated off the Earth's surface and atmosphere. A typical image derived from an infrared passive sensor consists of small equal areas referred to as pixels (7) arranged in regular rows and columns.

Each pixel has a numerical value called a digital number (DN) that records the intensity of electromagnetic energy measured for the area of ground represented by the pixel. The DN range from 0 to some higher number on a gray-scale. Each pixel is also given x and y coordinates to place it. The image can therefore be described in strictly numeric terms on a three-coordinate system with x and y locating the Pixel and z giving the DN displayed as a gray scale intensity value.

Passive sensors are described in terms of their spatial, spectral, and temporal resolutions. The spatial resolution of a sensor is the smallest area that is recorded as a separate unit (pixel). For instance, one-meter spatial resolution means that one pixel of a digital image represents an area on the Earth's surface measuring one meter in length by one meter in width. Spectral resolution refers to the number and dimension of bands (or wavelengths) of the electromagnetic spectrum that a sensor records. The higher the number of bands, the greater the sensor's ability to distinguish between objects. Temporal resolution, also known as repeat time, is the frequency with which a sensor passes over the same area.

Active Remote Sensing

Active remote sensing devices, on the other hand, emit high-energy electromagnetic radiation and record the relative amount and pattern of the energy that is reflected back. Many of these devices operate at wavelengths that not only penetrate cloud cover, but also vegetative cover and soil surfaces. The tradeoff for greater imaging capabilities, however, is increased complexity in data interpretation, as compared to passive sensor data interpretation.

Data Processing

After the satellite records the data, it is transmitted to a ground station for calibration and storage. The data may undergo various levels of processing before it is made available to the user. These levels range from simply correcting for transmission errors to performing advanced correction and analysis with model algorithms, depending on the needs of the scientists or user.

Once the data has undergone initial processing techniques, users may apply it for various purposes, from the simple production of an enhanced image to the more complex creation of image maps, thematic maps, and spatial databases. The data may also be used to develop statistical observations and graphs of the observed phenomena. To create maps and spatial databases, the initial data must be combined with other spatial data. An effective method to analyse the remote sensing data with reference to other spatial data is in a geographic information system (GIS).

Remote Sensing Data Integration with Geographic Information Systems (GIS)

Geographic information systems (GIS) are defined as computer systems capable of assembling, storing, manipulating, and displaying geographically referenced information (i.e. data points identified with respect to their location). GIS store information about the world as a collection of thematic layers that can be linked together by geography.

Remote sensing data applications and GIS have an established history of interdependency. GIS provides a format to distribute remote sensing data and to derive useable information from the data. Remotely-sensed data is also a critical means to create base GIS maps and update many data layers in the GIS. The integration of remotely-sensed data and GIS is particularly attractive because 1) the conversion of remotely-sensed raster-format data to GIS vector-format data is inexpensive and 2) remote sensing data offers a cost-effective way to visualize large geographic areas in a digital format.

There are two defining features of all GIS: the ability to overlay spatial data and the ability to change as new data becomes available. The first key feature of GIS programs is the capability to overlay multiple sets of databases into a map format that graphically explains the relationships between the data. Spatial data (points, boundaries, and lines) comprise the base of the map and can be supplemented with tabular data and image data (such as that from satellites). This powerful and versatile concept has proven invaluable for solving many real-world problems, from recording details of land use planning applications to modeling global atmospheric circulation cycles. The second key feature of GIS is their status as "dynamic maps" that can be updated and altered as needed. These maps may also be manipulated to perform scientific analyses and to create models of different environments.

In a simplistic example of GIS application, a map of city streets could be combined with latitude/longitude-referenced traffic flow data to create a map that reveals areas of frequent accident occurrence, potential detour routes, and even alternatives to improve traffic routing and alleviate rush hour stress. The same base map also may be reused to show, for example, changes in traffic patterns across time.

URBAN TEXTURAL ANALYSIS FROM REMOTE SENSOR DATA

Despite the new generation of very high spatial resolution sensor data (IKONOS from 1999 and QuickBird from 2001), predicted improvements in classification accuracy of urban land covers (and subsequent inference of urban land use) have yet to materialize substantially. Much of the obstruction to quality information extraction is still due to the traditional limitations of classifying image data representing urban areas: the high spatial arrangement of complex urban features and how to configure multispectral responses from land cover features into organized urban land-use categories (Barr, Barnsley, and Steel 2004). When launched, the desired objective of high spatial resolution sensor data was for increased clarity of terrestrial features, especially urban objects, by reducing per-pixel spectral heterogeneity and thereby improving land cover identification.

Clarity is certainly more evident in these finer-scale data than those from preceding sensors, but paradoxically this greater level of detail is also translated into many more unique per-pixel spectral combinations. For example, the residential land-use category can now be defined from much wider spectral variations, representing minute compositional mixtures of urban land covers, such as roads, houses, grasses, trees, bare soil, shrubs, and swimming pools, each conceivably a different residential land-use category. Following on, another limitation for improved information extraction from high spatial resolution sensor data is the reliance on techniques using traditional per-pixel spectral differentiation. To us this seems counterintuitive and we would like to see more neighbourhood-related methods, using textural and spatial parameters when dealing with fine-resolution image data.

Where traditional spectral approaches are designed to identify homogeneous features regardless of shape, textural and spatial algorithms measure both the variance within and the geometric configuration of whole urban objects, respectively. As a contribution to the growing literature, we outline an object-based pattern recognition technique that accommodates the concept of lacunarity for characterizing the textural properties of urban land cover (and therefore inferring land use) from high spatial resolution image data. In doing so, we consolidate the utility of geometric models not only for image data but for all discrete and textural spatial representations (Zhao and Stough 2005). Indeed, the ability to characterize the shapes of individual and groups of objects is a rapid area of research in computational geometry and at the heart of the recent developments in object-based models in many geographic information system algorithms.

Recall that remote sensor data are composed of multispectral pixel vectors that represent geographical objects and their relative configuration. We strongly adhere to the paradigm that geometric patterns, such as lacunarity, are valuable precursors for functional processes; in our application, the texture and spatial orientation of land-cover patterns derived from remote sensor data are both forerunners for analysing land-use juxtaposition and dynamic urban processes.

Lacunarity Approach

The lacunarity of an object is the counterpart of its fractal dimension. Lacunarity methods for urban analysis, and indeed many other applications in geospatial research, have already been reported by a number of researchers. Essentially, lacunarity is related to the spatial distribution of gap or hole sizes. For low-lacunarity measurements, all gap sizes are the same and geometric objects are deemed homogeneous; conversely, for high-lacunarity gap sizes are variable and objects are therefore heterogeneous. In other words, the variance or texture of gap sizes within the spatial delineation of geometric objects determines the level of lacunarity. Of course, textures that are homogeneous at small scales can be quite heterogeneous at large scales, and vice versa; therefore, lacunarity can be considered a scale-dependent measure of texture. Methods for calculating the lacunarity of objects were first given, in general terms, by Mandelbrot (1983) and were later implemented by various computer algorithms.

Work by Myint and Lam (2005b) developed two modified lacunarity algorithms: the binary approach, which was first introduced by Plotnick, Gardner, and O'Neill (1993), and a gray-scale routine, initially devised by Voss (1986) and used to test the effectiveness of lacunarity on high spatial resolution sensor data. This same desire to extract urban objects from fine-scale sensor data also forms the basis of this study, where we examine modifications to a differential box counting algorithm, first formulated by Dong (2000b).

Our study will also introduce two different gliding box approaches. The first uses overlapping boxes, in which the gliding box moves to a pixel next to the previous position, and the second uses skipping boxes, in which the gliding box skips the entire coverage of the previous box before moving to the next position. As a background, and according to the gliding box algorithm proposed by Allain and Cloitre (1991), n(M,r) can be defined as the number of gliding boxes with radius rand mass M. The computations of lacunarity values are given by worked examples where the overlapping box method is demonstrated by a 4 x 4 image or local window, while a 6 x 6 image is used to illustrate the skipping box method. The 3 x 3 gliding box used in both is the base of the cube box and is always an odd number to allow the computed value to be assigned to a central cell.

A column with more than one cube box may be required to cover the maximum image intensity values by stacking cube boxes on top of each other.

The number of cube boxes required to cover the image intensity surface depends on the pixel values in the 3 x 3 gliding box. Intensity values using the example are calculated at the first, second, third, and fourth positions of the cube boxes (overlapping and skipping boxes).

The minimum and maximum pixel values are 7 and 18, respectively, at the first position of the gliding box. With a cube box of 3 x 3 x 3, these values fall in box number 3 (value of u) and 6 (value of v), respectively. The relative height of the column is then 6-3 + 1 = 4 (u-v + 1). In the same way, we can compute the required parameters for all positions of the cube boxes as follows: For the second position of the gliding box, u= 7, v= 1 (the relative height of the column is (u-v + 1) or 7-1 + 1 = 7). For the third position, u=6; v=5 (the relative height is 6-5 + 1 =2).

RESEARCH DESIGN

Data and Study Area

We applied our technique to an IKONOS sensor image data of 4m spatial resolution across all four of its channels: blue (0.45-0.52 [micro]m), green (0.52-0.60 [micro]m), red (0.63-0.69 [micro]m), and near infrared (0.76-0.90 [micro]m). The image represents the settlement of Norman, Oklahoma and was captured on March 20, 2000. Only a subset of this IKONOS sensor image (614 x 447 pixels) covering a central portion of the metropolitan area was used to identify three land cover/land-use classes: grassland, commercial, and residential—all three capable of being delineated using manual interpretation.

Local Window

At the onset, it is important to note that the characteristic scale (Lark 1996) is an important parameter to be considered for effective identification of land covers with different texture appearances. A characteristic scale is the minimum distance between two pixels that completely covers a texture. In our study, anything between 21 pixels (84 m) and 27 pixels (108 m) was considered large enough to cover textures that represent all of the classes, especially complex residential and commercial land uses derived from the IKONOS multispectral image. Hence, 9 x 9, 15 x 15, 21 x 21, 27 x 27, 33 x 33, and 39 x 39 local window sizes were used to determine the optimum scale with which to identify land-cover classes.

From a previous study (Myint and Lam 2005b) that was based on a lacunarity approach designed by Voss (1986), we demonstrated that smaller gliding boxes have more discriminatory power than larger gliding boxes. And as a result, a small 3 x 3 x 3 gliding box is used in this study and is the very basis for the chosen window sizes dimensionalized by a factor of 3. Hypothetically, a window size should be small enough to cover only single land-cover features but large enough to guarantee sufficient spatial/textural information for the characterization of land-cover types.

In that case, a mirror extension of (w-1)/2 pixels around the image is necessary before beginning computation. What happens is that the algorithm is designed to extend automatically the image with (w-1)/2 pixels all around if the selected window size is w. As such, the size of an extended image is the original image size + (window size-1). Mirror extension is designed to copy the second-last row or column and add the next-to-last row and column, respectively. Then copy the third-last row and column and add the next-to-the-second-last row and column, respectively, and so on depending on the number of rows and columns required for the extension.

This mirror extension procedure is demonstrated by a hypothetical image and is considered more effective than other widely used methods, including those adding zero, one, last row/column data, or the mean values in the extended areas. Our favoured approach is also more accurate than other interpolation methods (e.g., kriging, inverse distance) because it does not alter any values in the original image. By using the mirror extension, we obtain the same mean and standard deviation statistics within the extended area.

Training Samples

With knowledge of the local area and ground checks, training samples were carefully selected to represent the three land-cover/land-use classes: grassland, commercial, and residential—descriptive statistics of their brightness values are shown. It is generally suggested that the standard deviation of training data should be around, or even less than, 10% of the mean coefficient of variation to guarantee homogeneous distributions vital to traditional spectral-based classifiers. Compared with the standard deviation, which is expressed in absolute terms, the coefficient of variation is considered a more appropriate measure for the comparison of data distributions across variables.

The standard deviation of the commercial and residential land uses is 37.1% and 34.3% of their respective mean values. At this stage, it is important to remember that groups of adjacent pixels generally exhibit positive autocorrelation, or at least display a high probability of having similar digital numbers. In our calculations, the grassland (Moran's l = 0.49; Geary's C = 0.43) sample exhibited a positive autocorrelation, whereas the commercial (Moran's l = 0.16; Geary's C=0.83) and residential (Moran's l = 0.09; Geary's C = 0.91) samples displayed random spatial arrangements with no patterns of clustering or homogeneity.

Band Combinations and Classifications

Finally, lacunarity algorithms, based on both skipping and overlapping boxes, were applied to transform bands 2, 3, and 4 of the IKONOS image. In addition, a number of layer stacks involving both lacunarity-transformed bands and original bands were generated in the following combinations:

- three original bands,
- three lacunarity-transformed bands,
- combination of three original bands and three lacunarity-transformed bands,
- combination of three original and one lacunarity-transformed band, and
- combination of two original bands and two lacunarity-transformed bands.

All window sizes were applied to stacks of three original bands and three lacunarity-transformed bands, as well as the combination of three original bands and one lacunarity-transformed band. Regarding the three separate lacunarity-transformed bands, only the 27 x 27 window was used, primarily to observe whether satisfactory accuracy levels were attained without the original bands. In the case of the combination of the three original bands and one lacunarity-transformed band, the lacunarity-transformed band 3 was used to test whether one lacunarity-transformed band could improve the accuracy of the three original bands. Experiments were conducted with all window sizes using lacunarity-transformed band 3, as well as testing lacunarity-transformed band 4 with the original bands using the 27 x 27 window for significant differences between lacunarity-transformed band 3 and lacunarity-transformed band 4. In the case of the combination of two original bands and two lacunarity-transformed bands, band 3 and band 4 were tested using the 27 x 27 window. All code for the modified lacunarity approaches was written in the C + + programming language.

Reference Map for Error Matrix

When examining the effectiveness of different classification algorithms, "wall-to-wall" comparisons were used by checking every pixel in the image.

The selected area covered 274,458 pixels (447 x 614) representing all three urban classes: grassland, commercial, and residential. Reference classes were collected by manual interpretation, using sound local area knowledge and a thorough ground survey, and then digitally delineated with negligible positional error. Overall accuracies produced by the combination of three original bands and three lacunarity-transformed bands using the overlapping box algorithm based on the 9 x 9, 15 x 15, 21 x 21, 27 x 27, 33 x 33, and 39 x 39 window sizes were 59.36%, 61.72%, 61.86%, 61.01%, 59.94%, and 61.65%, respectively. They demonstrate that overall accuracy increases slightly with expanding window sizes, despite a slight decline when using the 33 x 33 window, and it is fairly reasonable to suggest that the 15 x 15 and 21 x 21 window sizes are effective for this analysis.

However, overall accuracies produced by the combination of three original bands and three lacunarity-transformed bands using the skipping box algorithm based on the 9 x 9, 15 x 15, 21 x 21, 27 x 27, 33 x 33, and 39 x 39 window sizes were 58.31%, 59.29%, 57.40%, 56.41%, 56.92%, and 58.31%, respectively. Different window sizes produced inconsistent overall accuracies, and hence it is difficult to determine the optimal window size. In comparison, the overlapping box algorithm outperforms the skipping box algorithm, and because the overall accuracy of the original bands is 58.38%, it can be concluded that lacunarity using the overlapping box algorithm improved the classification accuracy of the original bands. However, we feel that accuracy was not significantly improved.

Overall accuracies produced by the combination of three original bands and one lacunarity-transformed band (band 3) using the overlapping box algorithm based on the 9 x 9, 15 x 15, 21 x 21, 27 x 27, 33 x 33, and 39 x 39 window sizes were 58.49%, 60.74%, 60.94%, 60.43%, 60.29%, and 61.35%, respectively. From this, we can deduce that there is only a slight difference between adding the one lacunarity-transformed band (band 3) to the original bands and adding three lacunarity-transformed bands to the original bands. However, it was also found that the combination of three original bands and three lacunarity-transformed bands produced higher accuracies.

Overall accuracies generated by the combination of three original bands and one lacunarity-transformed band (band 4) using the overlapping box algorithm for the 15 x 15 and 27 x 27 window sizes (58.14% and 58.99%, respectively) were lower than the combination with lacunarity-transformed band 3. Overall accuracies produced by the combination of three original bands and one lacunarity-transformed band (band 3) for the skipping box algorithm using the 9 x 9, 15 x 15, 21 x 21, 27 x 27, 33 x 33, and 39 x 39 window sizes were 58.87%, 60.07%, 58.60%, 57.37%, 56.93%, and 57.69%, respectively. The results show that there is no difference between adding three lacunarity-transformed bands to the original bands or adding one lacunarity-transformed band (band 3) to the original bands.

This may be mainly because the skipping box algorithm is not particularly effective in identifying texture features from image data. As mentioned earlier, we also tested the lacunarity-transformed bands generated by the overlapping box approach, individually.

These texture bands on their own gave slightly lower accuracy (57.80%) than the individual original bands, which tends to imply that spatial information alone may not be effective in identifying land-use/land-cover classes (an area we intend to research further). We also examined the combination of two original bands (bands 3 and 4) and two lacunarity-transformed bands (again bands 3 and 4) using the overlapping box algorithm based on the 27 x 27 local window. The results from this approach produced an accuracy level as high as the combination of three original bands and three lacunarity-transformed bands. This is probably because IKONOS bands 2 and 3 are visible bands and therefore highly correlated.

Urban remote sensing is currently experiencing a paradigm shift away from spectral-only classification and toward the identification of urban objects using spatial metrics and neighbourhood-based pattern recognition. Most of the conceptual restructuring is underscored by the availability of very high spatial resolution imagery, but also by the practical necessity of generating base maps of urban land use. Our work reported in this chapter is a contribution to research on spatial and textural identification of urban objects. We explored the utility of lacunarity to measure urban heterogeneity where results suggest that the overlapping box approach is more effective than the skipping box alternative.

We documented that there is no significant difference between window sizes, except for the case of the 9 x 9 window, and it might be reasonable to conclude that the 15 x 15, 21 x 21, and 27 x 27 windows are the most effective sizes in our study. It was also found that the combination of two original bands (bands 3 and 4) and two lacunarity-transformed bands (bands 3 and 4) is as accurate as the combination of three original bands and three lacunarity-transformed bands.

The combination of three original bands and three lacunarity-transformed bands is only a slight improvement on the combination of three original bands and one lacunarity-transformed band (band 3). However, what is certain is that the original bands alone or the purely lacunarity-transformed bands are not effective for this type of land-use and land-cover mapping.

Although variable, our results are very much in line with other documented work on textural and spatial characterization of high spatial resolution sensor data of urban land cover. It seems metrics, be they indices of dispersion, contagion, fractal, or lacunarity, are highly sensitive to the initial image segmentation, and inextricably affected by site, time, and scene.

Another reason for small and variable improvements in urban representation is the conceptual gulf between the computational limitations of the discretized remote sensor data model and the heterogeneous and dynamic nature of the urban landscape.

Such a gulf may prove difficult to bridge immediately with either spectral or textural/spatial indices, and a more integrative approach may ultimately be more appropriate. However, for now our work on lacunarity demonstrates another contribution to the potential for spatial metrics to characterize the increasingly finer spatial resolution of remote sensor data. We hold firm to our belief that highly detailed, spatially heterogeneous urban land cover can be measured by algorithms that are equally sensitive to geometric fluctuations.

EFFECTS OF ATMOSPHERE

In satellite remote sensing of the earth, the sensors are looking through a layer of atmosphere separating the sensors from the Earth's surface being observed. Hence, it is essential to understand the effects of atmosphere on the electromagnetic radiation travelling from the Earth to the sensor through the atmosphere. The atmospheric constituents cause wavelength dependent absorption and scattering of radiation. These effects degrade the quality of images. Some of the atmospheric effects can be corrected before the images are subjected to further analysis and interpretation.

A consequence of atmospheric absorption is that certain wavelength bands in the electromagnetic spectrum are strongly absorbed and effectively blocked by the atmosphere. The wavelength regions in the electromagnetic spectrum usable for remote sensing are determined by their ability to penetrate atmosphere. These regions are known as the atmospheric transmission windows. Remote sensing systems are often designed to operate within one or more of the atmospheric windows.

These windows exist in the microwave region, some wavelength bands in the infrared, the entire visible region and part of the near ultraviolet regions. Although the atmosphere is practically transparent to x-rays and gamma rays, these radiations are not normally used in remote sensing of the earth.

MICROWAVE REMOTE SENSING

There are some remote sensing satellites which carry passive or active microwave sensors. The active sensors emit pulses of microwave radiation to illuminate the areas to be imaged. Images of the earth surface are formed by measuring the microwave energy scattered by the ground or sea back to the sensors. These satellites carry their own "flashlight" emitting microwaves to illuminate their targets. The images can thus be acquired day and night. Microwaves have an additional advantage as they can penetrate clouds. Images can be acquired even when there are clouds covering the earth surface.

A microwave imaging system which can produce high resolution image of the Earth is the synthetic aperture radar (SAR). The intensity in a SAR image depends on the amount of microwave backscattered by the target and received by the SAR antenna. Since the physical mechanisms responsible for this backscatter is different for microwave, compared to visible/infrared radiation, the interpretation of SAR images requires the knowledge of how microwaves interact with the targets.

ELECTROMAGNETIC RADIATION

Electromagnetic waves are energy transported through space in the form of periodic disturbances of electric and magnetic fields.

All electromagnetic waves travel through space at the same speed, $c = 2.99792458 \times 10^8$ m/s, commonly known as the speed of light. An electromagnetic wave is characterized by a frequency and a wavelength. These two quantities are related to the speed of light by the equation,

speed of light = frequency × wavelength

The frequency (and hence, the wavelength) of an electromagnetic wave depends on its source. There is a wide range of frequency encountered in our physical world, ranging from the low frequency of the electric waves generated by the power transmission lines to the very high frequency of the gamma rays originating from the atomic nuclei. This wide frequency range of electromagnetic waves constitute the Electromagnetic Spectrum.

The Electromagnetic Spectrum

The electromagnetic spectrum can be divided into several wavelength (frequency) regions, among which only a narrow band from about 400 to 700 nm is visible to the human eyes. Note that there is no sharp boundary between these regions.

Wavelength units: 1 mm = 1000 μm; 1 μm = 1000 nm.

- Radio Waves: 10 cm to 10 km wavelength.
- Microwaves: 1 mm to 1 m wavelength. The microwaves are further divided into different frequency (wavelength) bands: (1 GHz = 10^9 Hz)
  - P band: 0.3-1 GHz (30-100 cm)
  - L band: 1-2 GHz (15-30 cm)
  - S band: 2-4 GHz (7.5-15 cm)
  - C band: 4-8 GHz (3.8-7.5 cm)

  - o X band: 8-12.5 GHz (2.4-3.8 cm)
  - o Ku band: 12.5-18 GHz (1.7-2.4 cm)
  - o K band: 18-26.5 GHz (1.1-1.7 cm)
  - o Ka band: 26.5-40 GHz (0.75-1.1 cm)
- Infrared: 0.7 to 300 μm wavelength. This region is further divided into the following bands:
  - o Near Infrared (NIR): 0.7 to 1.5 μm.
  - o Short Wavelength Infrared (SWIR): 1.5 to 3 μm.
  - o Mid Wavelength Infrared (MWIR): 3 to 8 μm.
  - o Long Wanelength Infrared (LWIR): 8 to 15 μm.
  - o Far Infrared (FIR): longer than 15 μm.

The NIR and SWIR are also known as the Reflected Infrared, referring to the main infrared component of the solar radiation reflected from the earth's surface. The MWIR and LWIR are the Thermal Infrared.

- Visible Light: This narrow band of electromagnetic radiation extends from about 400 nm (violet) to about 700 nm (red). The various colour components of the visible spectrum fall roughly within the following wavelength regions:
  - o Red: 610-700 nm
  - o Orange: 590-610 nm
  - o Yellow: 570-590 nm
  - o Green: 500-570 nm
  - o Blue: 450-500 nm
  - o Indigo: 430-450 nm
  - o Violet: 400-430 nm
- Ultraviolet: 3 to 400 nm
- X-Rays and Gamma Rays.

Photons

According to quantum physics, the energy of an electromagnetic wave is quantized, i.e. it can only exist in discrete amount. The basic unit of energy for an electromagnetic wave is called a photon. The energy E of a photon is proportional to the wave frequency f,

$$E = \mathrm{h} f$$

where the constant of proportionality h is the Planck's Constant,

$\mathrm{h} = 6.626 \times 10^{-34}$ J s.

EFFECTS OF ATMOSPHERE

When electromagnetic radiation travels through the atmosphere, it may be absorbed or scattered by the constituent particles of the atmosphere.

Molecular absorption converts the radiation energy into excitation energy of the molecules. Scattering redistributes the energy of the incident beam to all directions. The overall effect is the removal of energy from the incident radiation. The various effects of absorption and scattering are outlined in the following sections.

Atmospheric Transmission Windows

Each type of molecule has its own set of absorption bands in various parts of the electromagnetic spectrum. As a result, only the wavelength regions outside the main absorption bands of the atmospheric gases can be used for remote sensing. These regions are known as the Atmospheric Transmission Windows. The wavelength bands used in remote sensing systems are usually designed to fall within these windows to minimize the atmospheric absorption effects. These windows are found in the visible, near-infrared, certain bands in thermal infrared and the microwave regions.

Effects of Atmospheric Absorption

Atmospheric absorption affects mainly the visible and infrared bands. Optical remote sensing depends on solar radiation as the source of illumination. Absorption reduces the solar radiance within the absorption bands of the atmospheric gases. The reflected radiance is also attenuated after passing through the atmosphere. This attenuation is wavelength dependent. Hence, atmospheric absorption will alter the apparent spectral signature of the target being observed.

Effects of Atmospheric Scattering

Atmospheric scatterring is important only in the visible and near infrared regions. Scattering of radiation by the constituent gases and aerosols in the atmosphere causes degradation of the remotely sensed images. Most noticeably, the solar radiation scattered by the atmosphere towards the sensor without first reaching the ground produces a hazy appearance of the image. This effect is particularly severe in the blue end of the visible spectrum due to the stronger Rayleigh Scattering for shorter wavelength radiation.

Furthermore, the light from a target outside the field of view of the sensor may be scattered into the field of view of the sensor. This effect is known as the adjacency effect. Near to the boundary between two regions of different brightness, the adjacency effect results in an increase in the apparent brightness of the darker region while the apparent brightness of the brighter region is reduced. Scattering also produces blurring of the targets in remotely sensed images due to spreading of the reflected radiation by scattering, resulting in a reduced resolution image.

Airborne Remote Sensing

In airborne remote sensing, downward or sideward looking sensors are mounted on an aircraft to obtain images of the earth's surface. An advantage of airborne remote sensing, compared to satellite remote sensing, is the capability of offering very high spatial resolution images (20 cm or less). The disadvantages are low coverage area and high cost per unit area of ground coverage. It is not cost-effective to map a large area using an airborne remote sensing system. Airborne remote sensing missions are often carried out as one-time operations, whereas earth observation satellites offer the possibility of continuous monitoring of the earth.

Analog aerial photography, videography, and digital photography are commonly used in airborne remote sensing. Synthetic Aperture Radar imaging is also carried out on airborne platforms.

Analog photography is capable of providing high spatial resolution. The interpretation of analog aerial photographs is usually done visually by experienced analysts. The photographs may be digitized using a scanning device for computer-assisted analysis.

Digital photography permits real-time transmission of the remotely sensed data to a ground station for immediate analysis. The digital images can be analysed and interpreted with the aid of a computer.

In spaceborne remote sensing, sensors are mounted on-board a spacecraft (space shuttle or satellite) orbiting the earth. At present, there are several remote sensing satellites providing imagery for research and operational applications. Spaceborne remote sensing provides the following advantages:

- Large area coverage;
- Frequent and repetitive coverage of an area of interest;
- Quantitative measurement of ground features using radiometrically calibrated sensors;
- Semiautomated computerised processing and analysis;
- Relatively lower cost per unit area of coverage.

Satellite imagery has a generally lower resolution compared to aerial photography. However, very high resolution imagery (up to 1-m resolution) is now commercially available to civilian users with the successful launch of the IKONOS-2 satellite in September 24, 1999.

Satellite Orbits

A satellite follows a generally elliptical orbit around the earth. The time taken to complete one revolution of the orbit is called the orbital period. The satellite traces out a path on the earth surface, called its ground track, as it moves across the sky. As the earth below is rotating, the satellite traces out a different path on the ground in each subsequent cycle. Remote sensing satellites are often launched into special orbits such that the satellite repeats its path after a fixed time interval. This time interval is called the repeat cycle of the satellite.

COMMERCIAL REMOTE SENSING

The commercial satellite remote sensing market was initiated in 1972, around the time that the launch of the first Landsat mission (then referred to as ERTS-1) was being discussed. The success of commercial weather and communications satellites led the U. S. to believe that a land-observing satellite would eventually be able to pay for itself as private markets for the data grew. While the commercialization of Landsat has not been viewed as a success, it paved the way for the growth of the data distribution and value-added product industry sectors.

Private companies launching satellites can profit both as data collectors, and as intermediaries between raw data providers and the ultimate consumers. However, few private companies have launched successful Earth observation satellites. Between 1993 and 2000, following the passage of the 1992 Land Remote Sensing Commercialization Act, NOAA issued only seventeen licenses for private commercial satellites. Of the first four companies to launch private satellites, the two successes have been Space Imaging, Inc. and Orbital Imaging Corp. (Orb Image).

Associated Geospatial Technologies

Image-based GIS and photogrammetry comprised 69% of the geospatial activities market in 2000, with mapping, civil government, environmental, transportation, and national/global security markets controlling the highest percentage of sales. The market leader, Environmental Systems Research Institute (ESRI), controlled nearly $300 million of the $845 million total worldwide GIS software market in 1999. ESRI products provide a broad variety of applications to industries ranging from telecommunications and engineering to humanitarian assistance and environmental conservation. Intergraph holds the second largest market share, $238.7 million, of the total industry software market. Intergraph is the industry leader in providing GIS services and products to the utilities and telecommunications sectors, and is also an industry leader in public safety, transportation, and mapping sectors.

Due to the significance of the GIS market, commercial satellite companies such as Space Imaging and OrbImage have formed business relationships with GIS providers, predominantly ESRI. These two growing industries have both been aided by advances in the integration of remote sensing imagery and geographic information systems. Satellite data and images provide geospatially-referenced data for inclusion into GIS layers and can be used to create digital elevation models or other applications for GIS. ESRI's software can ingest, enhance, and classify IKONOS imagery and utilize it just like any other data layer in a GIS analysis. The imagery can serve as an incredibly detailed basemap upon which other layers are laid, or it can be used as an up-to-date data source from which various land cover and elevation features are extracted to populate multiple GIS layers.

The ASPRS/NASA Ten-Year Industry Forecast Currently, NASA and the American Society for Photogrammetry & Remote Sensing (ASPRS) are conducting a 10-year market survey of the remote sensing industry as defined by the "Space Act Agreement" between NASA and ASPRS.

In the first phase of the study, the team determined the baseline forecast of the U.S. Remote Sensing Industry (RSI) and associated geospatial activities. Four key findings emerged from the first phase of the study:

(1) The U.S. Remote Sensing Industry in 2000 was valued at $2.2 billion and is expected to grow at an average of 10-15% per year over the next five years, 2010;
(2) Currently the photogrammetry market is the largest Geospatial Market in terms of sales. Research and development for remote sensing, however, is considerably larger than that for photogrammetry or image-based GIS;
(3) Across all three sectors, the most active commercial markets are mapping/geography, environment, civil government, national/global security, and transportation. Environmental applications were rated one of the top four applications in all three sectors; and
(4) For the government sector, mapping, earth natural science research, and natural resource management were found to be the three most important missions.

Obstacles to Industry Growth

Numerous obstacles block the full realization of the remote sensing market. The Carnegie Endowment for International Peace identifies four critical factors that will ultimately decide the size of the market: the extent of government interference, the cost of commercial imagery, the time the data takes to reach the consumer, and the ability of the market to educate and interest consumers. Workshops have further addressed the problems related to acceptance of remote sensing technologies.

Issues ranging from a lack of access to standardizing data sets to "disconnects" between data providers and data users have been recognized, but not yet solved. However, one of the most pivotal obstacles, which must be overcome before others can be addressed, is that of acceptance of satellite data in a courtroom.

FUNDAMENTALS OF REMOTE SENSING

The process of remote sensing involves the detection and measurement of radiation of different wavelengths reflected or emitted from distant objects or materials, by which they may be identified and categorized by class/type, substance, and spatial distribution. The background required for use of remote sensing tools may seem overwhelming at first. The decisive factor in the successful application of remote sensing data, however, need not be the technical sophistication of the user, but rather the suitability and precise use of the tool to obtain accurate and relevant data. A general grasp of the technical process that transforms electromagnetic energy into useful information can improve and expand the appropriate use of these tools. Nevertheless, depending on the application, social scientists wishing to work with remote sensing imagery would do well to partner with physical scientists with a deeper understanding of how the imagery represents physical processes on the ground.

In the broadest sense, remote sensing refers to information gathered by measuring and interpreting signals. To use perhaps the simplest analogy, the human body is constantly involved in a variety of remote sensing tasks. Hearing and vision are two obvious examples, involving the gathering and interpretation of sound and light waves, both in limited ranges of the entire sound and light spectra.

Active and *passive* remote sensing are used to describe the way sensors gather data. To illustrate the two types of sensors, consider a snapshot camera, a sensor that captures electromagnetic radiation in the visible spectrum. Outdoors in full daylight, a camera is ordinarily used as a passive sensor in that it receives reflected visible light from its surroundings and uses optics, a shutter, and film to create a lasting image. At night, on the other hand, when there is inadequate light for most cameras to capture a useable image, the camera may employ a flash. The flash is emitted from the camera and bounces off the object to illuminate it, just as an active (or radar) sensor sends a burst of energy towards its target and then receives the reflected radiation. A key factor in the choice between passive and active sensor is the relative strength of the potential signal each system must measure. For a passive system, the source of the signal is ultimately the sun, which emits electromagnetic radiation at its highest intensity between the ultraviolet and infrared ranges. In the radar wavelength ranges, however, sensors must provide a signal of sufficient intensity to travel to the earth, and return with enough strength to be distinguishable from the background "noise" from other sources.

Energy Source or Illumination (A)-A basic requirement for remote sensing is an energy source to illuminate or provides electromagnetic energy to the target of interest. For passive instruments, this is usually the sun; for so-called active instruments, the sensor itself emits an pulse of energy.

Radiation and the Atmosphere (B)-As the energy travels from its source to the target, it will come in contact with and interact with the atmosphere it passes through. This interaction also takes place a second time as the energy travels from the target to the sensor. Inevitably there is a certain degree of atmospheric scattering of radiation.

Interaction with the Target (C)-Once the energy makes its way to the target through the atmosphere, it interacts with the target depending on the properties of both the target and the radiation.

Recording of Energy by the Sensor (D)-After the energy has been scattered by, or emitted from the target, a sensor (remote-not in contact with the target) collects and records the electromagnetic radiation.

Transmission, Reception, and Processing (E)-The energy recorded by the sensor has to be transmitted, often in electronic form, to a receiving and processing station where the data are processed into an image (hardcopy or digital).

Interpretation and Analysis (F)-The processed image is interpreted, visually or digitally/electronically, to extract information about the target illuminated.

Application (G)-The final element of the remote sensing process is achieved by applying the information that has been extracted from the imagery about the target in order to better understand it, reveal some new information, or assist in solving a particular problem.

ELECTROMAGNETIC ENERGY

Remotely sensed data are collected in many regions of the electromagnetic spectrum. Data recorded from each part of the spectrum can provide distinct information on characteristics of the Earth's surface or properties of the atmosphere. For example, healthy green vegetation reflects highly in the near-infrared region of the spectrum, whereas water bodies tend to reflect only a small amount of incoming radiation in the visible region. All remote sensing instruments collect electromagnetic radiation that is reflected, emitted, or scattered from the Earth's surface and atmosphere. So-called active sensors such as radar and lidar emit energy that bounces off the land or water surface and returns to the sensor to be recorded. The way the energy is directed or scattered by the surface, and the time it takes for the energy to return, reveals information about surface characteristics. Because of the long wave lengths employed by radar, the signals can penetrate clouds, thereby allowing scientists to record information about normally cloud-covered areas. This is an asset in tropical areas such as the Amazon River basin.

Passive sensors, on the other hand, typically rely on solar illumination of the Earth's surface, though some are equipped to detect night-time lights and gas flares. These sensors are "passive" because they do not emit their own energy, but rather rely on energy reflected or emitted from the earth's surface. Unlike radar sensors, they are unable to penetrate clouds. It is interesting to note that the visible portion of the spectrum-those wavelengths that humans can see-is a very small segment of the spectrum. Part of the strength of remote sensing is that it enables scientists to "see" portions of the spectrum that are outside the range that the human eye can detect. Scientists can combine non-visible portions with visible ones through colour composites, assigning each *band* (or portion of the spectrum detected by the instrument) the colours red, green and blue.

A sensor's bandwidth and the number and placement of bands (within the spectrum) define its *spectral properties*. Panchromatic sensors measure reflected energy in a single portion of the electromagnetic spectrum, usually the visible to near-infrared regions. Multispectral sensors, on the other hand, collect reflectance information in discrete portions of the spectrum, with each being recorded as a separate image called a band or channel. When these bands are displayed on a computer, with one band shown through each of the blue, green and red channels of the monitor, they yield a combined colour image. Landsat 7's Enhanced Thematic Mapper, for example, is a multi-spectral instrument that collects data in eight bands-three visible (one each for blue, green, and red), a near-infrared, two middle-infrared bands, a thermal-infrared and a higher spatial resolution panchromatic band. By contrast, the Moderate Imaging Spectrometer (MODIS) collects data in 36 different spectral regions, and the Airborne Visible/Infrared Imaging Spectrometer (AVIRIS) is a hyperspectral instrument that collects data in 224 spectral bands.

Ultimately, what a sensor measures is the intensity of radiation that actually reaches the sensor, which is termed the *at-satellite radiance*. Radiance values are commonly translated into digital numbers (DNs). The possible range of DNs varies between sensors, although ranges of 0-255 (for 8 bit images) and 0-1023 (for 10 bit images) are common, with higher values corresponding to greater brightness. Radiance is captured by a two dimensional array of picture elements, or *pixels*. A DN for a pixel in a specific band is determined by the intensity of the radiance captured for that particular portion of the electromagnetic spectrum. If space-based passive sensors were able to accurately, precisely and repeatedly capture the actual reflectance from a feature on the ground, regardless of the time of day, season or weather conditions, much of the hard work of image processing would be eliminated. But the reality is that the atmosphere scatters radiation that is reflected back out to space. Smoke, haze, clouds and humidity exacerbate the problem, and can block reflected energy entirely. Data from shorter wavelengths are more likely to be blocked or scattered by clouds or atmospheric particles, whereas images using sensors capturing longer wavelengths are less likely to be disturbed by atmospheric conditions between the sensor and the target object.

GENERAL IDEA ABOUT ELECTROMAGNETIC SPECTRUM

The idea of the electromagnetic spectrum is one of those great scientific syntheses of observable natural phenomena that should be part of the knowledge of any scientifically educated person. Other similar syntheses are things like the kinetic-molecular theory of matter, the concept of the vastness of the Universe and our relation to it, the idea of deep time and the evolutionary development of life, the DNA code of life and the ideas of basic biochemistry.

Since the information technologist makes use of the electromagnetic spectrum a great deal, it is appropriate to devote a lecture to it in our course.

Electricity and Magnetism

These were two disparate ideas that were first discovered in antiquity: the Greeks gave the name (as in so many areas) through their term for amber. Amber is a naturally occurring fossizilized tree sap, made famous in the movie Jurassic Park. Anyway, the Greeks had observed that amber when rubbed could attract small pieces of dust or other light trash. Also, static electricity sparks were also known.

Naturally occurring magnetic material ("lodestone" or a type of iron oxide with a higher concentration of iron than usual) was also known to the ancients. It is said that the Chinese were aware that a piece of lodestone could be floated on a piece of cork, and would stabilize to a consistent North-South orientation, thus making possible a compass.

Hewever, it wasn't until the 1700s that the beginnings of our modern "scientific" approach to electricity and magnetism began.

Alessandro Volta noted a connection between electricity and life when he made dissected frogs legs twitch by touching them with dissimilar metals. [This inspired Mary Shelley, the wife of the famed poet Percy Bysscche Shelley, to write a famous novel, whose theme was "There are things that Man is not meant to know..."] Anyway, Volta had been experimenting with placing dissimilar metals, *e.g.* copper and zinc, into salt solutions, and obtaining an electric current.

It was early shown that if wire was wrapped into a coil, and connected to a battery, that a magnet was produced in the area close to the coil. If a piece of iron was placed inside the coil, the field was greatly magnified, giving a device called an electromagnet. If the iron was a hard steel alloy, it would retain the magnetism after the current was removed, allowing the manufacture of permanent magnets. Lodestones were quite weak compared to artificial magnets.

Hans Christian Oersted observed that if a wire was wrapped around a compass, the compass would deflect.

This was a unique phenomenon. The uniqueness came from the fact that the interaction was at right angles. The compass needle didn't line itself up with the wire, but at right angles to it. This was the first "vector cross product" relationship discovered. All other interactions hitherto known acted in a straight line between centers: gravitational attraction, electric force and magnetism.

Scientists had been experimenting with pure electric force also. To observe pure electric force out in the open requires charges of thousands of volts, which, fortunately, are easily obtained by friction. When you give someone a shock by "static electricity" you are using thousands of volts. It is not lethal because the amount of electric current is small.

[I use the term "scientists" here. In those days, in the 18th and 19th centuries, the usual term was "philosophers" or "natural philosophers". Our "science" was their "natural philosophy."]

[One can make water analogies of these ideas; in fact any electrical device can have a water analogy. One can think of electric charge as being a quantity of water, current as the rate of flow of water, and voltage as the force or pressure propelling the water. The force or pressure comes from a source of energy; either the water must be pumped up to a height to give it pressure-that is, the water has potential energy, or else a pump must take energy from some other source to force the water to move directly. Our city water system uses a combination of both methods. Strictly speaking, voltage is called potential difference, and represents the potential energy per unit charge, measured between two points in space.]

These studies were being carried out mainly in Europe, especially France and Italy. England was mainly preoccupied with the developing industrial revolution, and its scientific workers were clarifying ideas about energy and thermodynamics, motivated by the development of the steam engine.

America was regarded by European intellectuals as a land of wild Indians and stagecoaches, if they thought about it at all. The idea that someone in America could do Natural Philosophy seemed absurd. However, Benjamin Franklin in the middle of the 1700s had a brief period in his life when he had made enough money from his printing shop to semi retire, but hadn't become embroiled in the American Revolution as representative in France yet.

He experimented with electricity, leading up to his famous kite experiment in which he showed that lightning was the same basic thing as static electricity created in the lab. Since he knew that lab static dissipated quickly if the charged object had sharp points on it, he thought that maybe lightning could be dissipated before the actual stroke if sharp objects electrically connected to the earth were to be placed on high roofs. This was shown to be effective. Some feel that this vivid demonstration of the ability of man to evidently interfere with the "will of God" successfully was a significant force leading to the growth of secularism in Christendom. There was, after all, no more vivid demonstration of the "will of God" than to be struck down by a lightning bolt!

A similar formula was shown to work for idealized magnetic poles:

The constant mu zero is used to make the numbers work out. M1 and M2 are the magnetic pole strengths. [In fact this equation is an idealization more so than the electric equation above. An isolated magnetic pole is an unachievable abstraction-in practice magnets always exist as dipoles-N and S poles fairly close together. Electric poles can in principle be separated as much as needed.] By the middle 1800s quite a bit of electrical technology had been worked out. Things like telegraph lines were beginning to appear, allowing instant communication over vast distances. In fact, the mid 19th century saw communication finally surpass that under the Roman Empire 2000 years earlier!

Electric batteries, motors, generators existed. Electric power stations were starting to appear.

James Clerk Maxwell studied all the disparate physics and engineering involved in this field, and formulated a theoretical formulation that tied it all togther. This led to the great triumph of "field theory" in which each point in space could be thought of as posessing a magnetic and electric force; that is, there existed an electric and a magnetic field.

Oersted had shown that these were related, and the electric motor and generator were the practical embodiment of this discovery. Physics students learn about these when they study electromagnetism in 2nd or 3rd year. A good part of the early technical education of a would-be physicist is to learn the necessary mathematical tools, that is, integral and differential calculus, to enable manipulation of these equations. A physicist has to learn how to apply these to specific real-world situations (albeit a bit idealized for pencil and paper calculations). This theory was a revolution in the 1870s when it appeared; it took a generation for it to be absorbed into the technical consciousness of physicists.

When undergraduate physics students first start to work with these equations, there is an understandable "rush" in their minds; all these amazing things revealed by such a small number of symbols. To me, an interesting issue is, Where does the information actually lie? After all, it takes an educational process spanning many years to enable the physicist to work with these equations; is the information in the symbols, or are the symbols a bit like very high level function calls in an API environment, in which they trigger a complex series of mental events that yield the result. The function call without its substrate of lower level operations would be useless. Maxwell's equations without the technical education are just so many hieroglyphics, as they may well seem to CPSC students!

Enter the Electromagnetic Spectrum

Maxwell initially formulated his theory based on an idea of the electromagnetic effect being due to a sort of fluid flow. From the basic equations, he worked out the speed of the interaction; this is a calculation including the two basic physical constants: the electric constant and the magnetic constant. Quantitatively, the velocity comes out to 3 E8 m/s, or the speed of light. This immediately led to the idea that some sort of "electromagnetic" waves could exist, and the possibility that light might be such a wave.

Attempts to detect such waves were carried out by Heinrich Hertz, who demonstrated the first apparatus for creating and detecting low frequency electromagnetic waves, *i.e.* "radio".

In 1902, Guglielmo Marconi demonstrated the transmission of radio waves across the Atlantic. This was the beginning of the radio industry. One of the main applications was the implementation of ship to shore telegraph. Previously, a ship was completely out of touch with the world while on its voyage. The sinking of the Titanic in 1912 was an early example of the use of this technology to summon help.

Gradually, engineers learned how better to control radio waves, allowing for "wave bands" or fixed frequency "bands" allowing simultaneous use of electromagnetic waves by different people in the same area for different transmissions. Today, of course, we have this with our radio stations; we "tune them in" by adjusting our receiver to the desired band.[As a side issue, what about the idea of a fluid for electromagnetism? This idea persisted for some decades, and was finally put to rest by the Michelson-Morley experiment, showing that no physical medium existed (or if it did, it carefully moved so as to nullify any attempt to measure the movement of any apparatus through it!) This led to Einstein's Special Theory of Relativity, and the equivalence of mass and energy, and the atomic bomb.]

Frequency Bands

Early radio technology was of necessity very simple minded, one might say crude. Although high frequencies could carry more information the simple electronic valves (*i.e.* "tubes") could only operate at relatively low frequencies, in the megahertz region. Thus, the first broadcast channels appeared around 1 MHz.

A channel can be thought of as an independent path for information. Modulation is the process of coding information into a channel. A technique called "amplitude modulation" or AM allowed audio information to be coded into a radio frequency channel of narrow bandwidth, in fact only double the frequency range of the audio itself. Since audio goes to 20KHz, this would imply that an AM channel of 40 KHz width would be needed. In fact, it was decided to use a bandwidth of only 10 KHz, since this was adequate for the sound reproduction technology back in the 1920s when radio appeared. [Radio killed off much live theater, vaudeville, and the domestic sale of pianos!] Thus, we had about 100 possible channels in the range 600-1600KHz. A problem with all electromagnetic channels is that in any one reception area, only one channel at any one frequency is usable. Because AM radio stations "carry" quite a long way-the lower frequency ones diffract around the earth's curvature, the higher frequency ones bounce off the ionosphere, especially at night, the 100 channels must be spread around the country fairly thinly. If two channels on the same band are picked up at the same time by a receiver, a squealing noise is heard.

Above the AM band, a number of miscellaneous AM bands developed for various services, collectively called "short wave". The ionospheric bounce was particularly effective in some of these bands, making them useful for military transmissions, and propaganda broadcasts, especially in the "Cold War" from 1950 to 1990. I can remember as a teenager playing with shortwave tuning into Russian rantings (in English) about the "dictatorship of the Proletariat". I can't imagine who would want to listen to this!.

Relatively recently, a band at about 27MHz was allocated for AM Citizen's Band radio. This originally had 24 channels, later expanded to 40. You can see how narrow the CB band looks on the diagram, compared to AM radio. This is due to the logarithmic plot I have used for frequency in the horizontal axis. More modern technology has allowed these higher frequencies, where there is more room for channels, to be used economically, with the aid of "crystal filters" and "frequency synthesis" techniques, not available in the 1920s.

An inventor named E H Armstrong invented many of the basic analog circuits used for radio receivers. A problem with AM radio was its lack of noise rejection. He conceived of the use of another technique, frequency modulation or FM. This would need about 10 times the bandwidth per channel, requring the use of higher frequency bands where there would be enough Hz available. He worked with RCA, who gave him the use of a laboratory and a transmitter setup on the roof of one of the new New York skyscrapers then being built. However, as mentioned in a previous page, Sarnoff, the head of RCA, decided to throw his resources into the development of television. Thus, FM lost the support of RCA, Armstrong committed suicide, and TV ended up with much of the spectrum space. You can see FM squeezed in between the TV channels. RCA threw Armstrong a bone by making the TV audio frequency modulated, but the picture is AM.

Each FM band gets 100KHz, which Armstrong felt was a sell-out; he wanted more. Each TV band takes about 4500Khz, with channels about 6000Khz apart. This separation allowed cheap TV receivers to reliably receive adjacent channels in the same market area.

The long carry of short wave stops working at around 20MHz, so that TV and FM stations are basically "line of sight." This necessitates high transmitting antennas, and allows each major market to have lots of available channels.

In effect, a copy of the electromagnetic spectrum can be carried inside a cable, as an electrical analog, up to frequencies of about 500MHz or so. So, we have cable TV and other services, compatible with the airborne electromagnetic channels.

Finally, at 800 MHz, we find the original cell phone band, used for analog phones.

The next image shows the two digital cell phone bands. I need to find out what bands WI-FI uses, and the upcoming ultra high speed wireless network bands. This latter, I understand, uses "spread spectrum" technology; the information is carried using many different frequencies rather than on single frequencies. This makes interference a serious problem, except that the power is kept very low, reducing range.

Historically, the TV and Cell bands were originally used only for RADAR (Radio Direction and Ranging). This was developed as a crash programme during World War 2. Like many technologies, it was invented by the British but developed elsewhere, in this case the USA. The British invented the "cavity magnetron" vacuum tube, in which high voltage electrons are injected into a very strong magnetic field, causing them to swirl in a curved path, oscillating at the desired gigahertz frequencies. [When this was first demonstrated to the Americans, one of the leading scientists, I I Rabi, said "It's easy-an analog of a whistle" prompting one of the others there to remark sourly "OK, Rabi, how does a whistle work?"] In fact many audio devices and microwave devices are analogous, since the wavelengths are similar. Treble frequencies for audio have mechanical waves similar in length to gigahertz electromagnetic microwaves.

Radar was very crude in its use of the frequencies, basically just an on-off setup. Our modern cell phones are far more sophisticated in their modulation techniques, allowing thousands of channels in any local area.

Apart from crude things like microwave ovens and heat lamps, a huge stretch of the electromagnetic spectrum is currently unused. One reason is that our atmosphere is opaque to these frequencies. This in fact is a cause of the greenhouse effect. The short wave infrared can penetrate the atmosphere, coming from the Sun. It is absorbed by the earth, and re-emitted as long wave infrared. This is because the earth is much cooler than the Sun, and the wavelength produced by an emitter of electromagnetic radiation shortens as its temperature rises.

We generate radio by electrical excitation. Heat will also generate it, although not in a useful fashion, as there is a mess of random frequencies produced. Frequencies like our TV would be characteristic of temperatures very close to absolute zero.

Finally, we reach the visible band, a very narrow strip from about 400 to 800 THz. This band exists because our atmosphere is transparent in this narrow range, and our eyes thus work in this area. Fortunately, the air blocks ultraviolet except the very near UV. This is fortunate because this is the start of the bands that can disrupt chemical bonds.

We are starting to make use of light in CPSC, for fiber optics. Certainly, there are a lot of Hz there. However, our modulation is still rather crude, although it is improving; in fact, unexpected improvements may have contributed to the great internet "bust" in 2001. Companies laid huge amounts of long distance fiber, expecting that each fiber could carry maybe 5 channels. So, methods of sending 200 channels appear, resulting in a vast oversupply of fiber bandwidth! Channels can be sent over fiber using different colors. This is called "wave division multiplexing". I suppose our AM and TV channels could rejoice in the same term!

Visible light in many ways is a "watershed" in terms of wavelength and frequency. The visible band shows both particle and wave characteristics, depending on what experiment is done to observe. Visible light shows wave effects like diffraction and interference, but also shows particle like effects (*e.g.* the photoelectric effect; Einstein got his Nobel from his explanation of this effect, not from relativity).

A whole treatise could be written on the physical causes of visible colors. Basically, they result from energy transitions within atoms, of electrons moving from one state to another. Colors towards violet are caused by higher energy transitions. As we move into the UV, the transitions can involve molecular disruption. The Ozone Layer blocks near UV; if it gets through the degree of molecular disruption on the surfaces of living things becomes an issue, with skin cancer, etc, possibly becoming a serious problem.

Continuing to higher frequencies, X rays are characteristic of very high energy electron transitions. They are produced in CRTs by the electrons slamming with 20000 Volt potential into the inner face of the tube. The tubes are made with lead glass to block these X rays. "Harder" X-rays come from low level nuclear disruption; the nucleus of the atom is about 6 orders of magnitude smaller than the atom itself, and about 6-7 orders of magnitude harder to change in state. Nuclear changes of state (*e.g.* atomic or hydrogen bombs, or the core of the sun) release gamma rays.

There are isolated extremely high energy gamma rays, about 10 orders of magnitude beyond my diagram, occasionally seen. Their source has been a mystery for decades; I believe some tentative explanations now exist.

Shannon's Law and Data Transfer

The Bell Telephone System ran the telephone network in much of the USA until the late 1970s or early 1980s as a government regulated monopoly-almost a public utility. Since they had a protected position, with a guaranteed profit, they could engage in long term research into issues related to communications.

Since by law they could not sell ancillary discoveries for profit, they basically gave away many of their great developments, such as the transistor. [We might note that their Unix developed just as the Bell system was being broken up by the US Government, and the legal restrictions were also being removed. So, Bell started trying to sell Unix licenses, for an amount about equal to the gross national product of Luxembourg, with the predictable result that several non Unix Unices developed, thus "forking" the development of the system, producing several incompatible versions, and thus aiding Microsoft in its march to hegemony.]

1948 was a great year. The transistor was announced to the world, the Manchester University (England) "Baby" electronic computer ran the first stored programme in June, and Claude Shannon announced his communication theory.

Shannon worked much of his professional life at Bell Labs. His main claim to fame is his study into the error-free information capacity of an analog channel. (We mustn't forget as we think about digital technology that it all is really analog underneath). Analog operation involves continuous functions of variables such as voltage, current, time.

Noise is any unwanted energy; it arises from many causes. The most basic is thermal noise caused by the random motion of atoms and molecules in all matter. Noise comes in from the cosmos also; some of the "snow" visible on a TV with an antenna if tuned to a vacant channel may be from this source [Some of this may be echoes of the Big Bang!]. Not surprisingly, human technology produces lots of noise; any electrical switching or sparking produces electromagnetic noise. Computers are of course changing state billions of times per second; each state change is physically done by voltages within the system changing. This will propagate electromagnetic noise. [In fact, there are regulations regarding the amount of RFI (Radio Frequency Interference, or noise) that can be produced by any electronic device. One problem with all this sort of regulation is that compliance is expensive; a small company must send its device to a consultant lab which will run the necessary tests and sign the forms that go to the regulators, a process that may cost $10,000 to $100,000. This process is especially onerous in Europe; where some have cynically suggested that one reason for the stiff regulations is that the big industries, like Siemens and Philips, can maintain staffs for dealing with compliance issues, while small companies face a serious barrier to entry in the marketplace as a result. In North America, the regulations are not so stiff, and in Asia, it is still the "Wild West" in many places.]

Signal to noise ratio is the ratio of the signal value (voltage, current, energy, pressure, or whatever) divided by that of the noise. I will use a value of 1000 in the examples below. This corresponds roughly to AM radio, or the telephone on a good day. High fidelity audio amplifiers can achieve values of around 100,000, limited by thermal noise in the transistors and resistors. [Professionals often measure this in decibels. For voltage or current measurements, the decibel (dB) value can be determined by multiplying the log (10) of the S/N by 20, so a S/N of 1000 would be 20x3 or about 60dB, and 100,000 would be about 100dB.]

Finally, let's consider light. The visible spectrum is about 4 x 10^14 Hz wide. If we use the same 1000 S/N we get a gross data capacity of 4 x 10^15 bits/sec, or about 4 x 10^14 bytes/sec, or about 4 x 10^8 Megabytes per second, or about 4 x 10^5 gigabytes per second, or about 4 x 10^4 DVD movies per second!

Unfortunately, this assumes that we can control light the way we can control radio frequencies, with detailed control over each electromagetic "vibration". This is not possible with light due to things like "Doppler broadening" caused by the thermal motion of the atome producting the light. However, much progress can be made from our current technology.

PLATFORMS AND ORBITS

High-altitude remote sensing originated in the mid-1800s with aerial photography by balloon and, in at least one instance, the use of cameras attached to the underside of birds. Airplanes became the dominant remote sensing "platform" by the early 20th century. This practice continues to evolve for wartime, intelligence, commercial and government applications. An advantage of airborne remote sensing, is the capability of offering very high spatial resolution images (20 cm or less). The disadvantages are low coverage area and high cost per unit area of ground coverage. It is not cost-effective to map a large area using an airborne remote sensing system. Airborne remote sensing missions are often carried out as one-time operations, whereas earth observation satellites offer the possibility of continuous monitoring of the earth. The development of satellite remote sensing had greatly improved the ability to cover large areas.

There are two groups of satellites depending on the orbit in which they are placed. A *geostationary* orbit is established when a satellite is placed at a very high altitude, roughly 36,000 km above the earth's equator, and caused to orbit with the earth's rotation (called a *prograde* orbit). The altitude may vary slightly from one geostationary satellite to the next, depending upon the mass of the satellite, but, for the most part, this is a fundamental physical constraint. The rules of geometry-that is, the sight line from the satellite's position above the equator to the farthest edge of the earth's sphere-dictate that geostationary satellites can only "see" limited area at any one time. The laws of physics and capabilities of engineering limit their spatial resolution to a range of about 1 to 10 square kilometres. The total image or scene size, known as the *field of view*, is often thousands of kilometres. Thus, unless a geostationary satellite spins or turns its optics, its view is necessarily fixed. This allows for continuous monitoring, and often a very large, synoptic view of much of one entire hemisphere.

The coarse (km range) resolution versus the wide, continuous field of view constitute the main tradeoffs to consider for this orbit type. These characteristics make geostationary satellites best for collecting weather and climate data (such as cloud cover and surface temperature) and relaying communications data, although AVHRR data are used for global-and regional-scale land cover analyses.

The other group of satellites, by far the largest group of earth-orbiting satellites is with the *sun-synchronous* or *polar-orbiting*. These are lunched below the altitude of geostationary satellites closer to the earth's surface, at orbits ranging from 700km to 1000km. These satellites usually orbit at a steep inclination relative to the equator, in the direction opposite the earth's rotation, known as a *retrograde* orbit. When the satellite's orbit and earth's rotation are combined, they result in an s-shaped path relative to a map of the earth's surface. Given enough time, the orbits and rotations of the earth bring the satellite over the same location, leading to the term *exact repeat* satellites. The number of orbits between each return to the same longitude and latitude is called the *repeat cycle*. These satellites usually orbit the earth in roughly 100 to 120 minutes, circling several times per day, returning a satellite to the same position over the earth's surface only after 2 weeks or more. The speed of motion limits the time that a satellite spends over a location, and the amount of time a scanner can "look" at any single ground cell (called the *dwell time*). Most exact repeat satellites that use passive sensors are also in sun synchronous orbits, meaning that they cross the same latitude at the same daylight time with each orbit, but with their location shifted to a different longitude. Their lower altitude allows these satellites to obtain images with spatial resolution ranging from 1-200 meters per side of a pixel, and an image width ranging from tens to thousands of kilometres per scene.

As the satellite passes over the earth's surface, its motion can be described in terms of a ground track that it follows at a certain altitude. Most satellites are *nadir looking*, meaning that their sensing equipment is aimed straight down toward the center of the earth. However, normal measurements generally include areas substantially on either side of this ground track, and that total area is called the *swath width*. Because most low-orbiting satellites follow a polar orbit, their ground tracks, and thus their swaths, are spread furthest apart at the equator, and are compressed at the poles. As a consequence, there is an overlap, called *side lap*, of neighboring swaths. Logically, this side lap is smallest at the equator and increased at the poles.

A basic understanding of ground tracks, swath widths, and side lap is helpful in designing a remote sensing experiment for a few reasons. First, weather and other temporary conditions may prevent good data acquisition for the first and/or "best" pass of a satellite directly over a given location. In this case, side lap may allow for multiple acquisitions of the same location on the surface with only a short delay, provided that the target location can be seen during two or more subsequent passes of the same satellite. This situation is more likely to occur if the sensing target is at high or low latitude than at the equator. Also, side lap may allow experimenters to obtain data for a single location in much more rapid time series than would be possible if the experimenters were to wait for the satellite to exactly repeat its path-a matter of hours, rather than weeks. Finally, side lap may allow the same location to be viewed from slightly different angles at slightly different times of the same day with neighboring orbital paths. This may provide additional information from shading caused by the sun's angle, and other factors that change with relatively small differences in position and time.

SENSORS CHARACTERISTICS

A sensor is characterized by its spectral properties (number and placement of bands), its orbital altitude and path, its swath width, and its *spatial resolution*. Spatial resolution is measured in terms of the size of one pixel projected on the ground. Spatial resolution is directly tied to the size of the features that can be resolved (or "seen") on the ground. The higher the resolution, the less likely that there will be "mixed pixels" in which radiances effectively represent an average of land cover types in the ground area represented by that pixel (e.g., half lake and half forest). Commercial high resolution sensors have a spatial resolution in the 0.6-10 meter range, medium resolution sensors fall in the 10-

50 meter range, and low resolution sensors have greater than 50 meter resolution.

Until the advent of the commercial satellites IKONOS and QuickBird, with resolutions of one square meter or finer, high resolution imagery was the exclusive province of intelligence-gathering agencies. Most social science applications do not command the financial resources required to obtain such high resolution data, nor are images of this resolution generally required, except perhaps in the area of international relations, law and policy. Most social science research tends to utilize polar-orbiting satellites with medium spatial resolution, such as the Landsat, SPOT, TERRA and, more recently, AQUA satellites. They provide good spectral and ground resolution, with multiple visible, infrared, and panchromatic bands and pixel width ranging from 5 to 30 meters.

DATA PROCESSING, INTERPRETATION AND ANALYSIS

Much of the technical work of remote sensing involves pre-processing and applying radiometric and geometric corrections to imagery to compensate for errors due to factors such as atmospheric interference of incoming radiation and sensor and data stream irregularities. Once such corrections are applied, imagery must be *georeferenced*. Georeferencing is the process of taking the image in its raw format (rows and columns of data) and linking it to the land that it covers. Images are georeferenced by linking spatially distributed control points in the satellite image to points on base maps or points referenced in the field through global positioning systems. The raster data in the image is thereby registered to a Cartesian coordinate system, and can be combined with other georeferenced data sets in a geographic information system.

The processed data can now either be visually interpreted or classified using manual or automated processes. The main elements of visual image interpretation involve gradients of tone or colour, resolution, size and shape, texture and pattern, site and association, and height and shadows. Given their knowledge of the characteristic spectral signatures of different land cover types, scientists my inspect black and white images of each band separately in order to identify features and patterns.

For many purposes, data that is collected from the earth's surface, which represents a continuous variation, needs to be categorized (de Sherbinin *et al.* 2002). Image classification is the process of creating discrete classes or categories of land cover, utilizing information from some or all of the bands to group together pixels with similar spectral signatures. Supervised classification entails providing the software with sample pixels that represent specific features, such as boreal forest, and then having the computer classify every pixel with a similar *spectral signature* as boreal forest. Analysts may also use images from different seasons in order to discriminate vegetation cover types that have different phenologies, such as deciduous and evergreen forests. In unsupervised classification, the analyst specifies the desired number of classes, and the computer automatically sorts the pixels according to their spectral signatures. The analyst then labels the resulting groups based on some local knowledge of the land cover patterns.

Once classified, it is necessary to verify or validate that the output product accurately represents the actual composition, content, structure or land surface characteristics being mapped. Validation requires either field visits, ground-truthing or comparing the classified image with existing maps or images of sufficient detail. Statistics can be derived for the classified imagery indicating the general and specific (class-wise) agreement between the pixels or classes used, letting the user know which were classified correctly and which ones were not.

Validation results are also sometimes presented as a percentage value associated with the map that communicates how accurate the map is on a per pixel basis. Since the highest confidence rankings reported by satellite land cover data sets are between 85% and 90% (for the easiest types of land cover to classify), for an image with a per pixel accuracy of 85% the probability that one pixel out of four is incorrectly classified is close to 0.50. The output of remote sensing data analysis can be presented in a variety of ways including a printout of the enhanced image itself a spatial database, summary statistics and/or graphs (Jensen 1996). The output data can be integrated with a geographic information system (GIS) database for further analysis.

7

Remote Sensing Satellites

UNITED STATES GOVERNMENT REMOTE SENSING PROGRAMS

The United States began the current phase of Earth observation from space with the launch of the first Landsat satellite (ERTS-1/Landsat-1) in 1972. Currently, the United States has ten Earth observing satellites in orbit. Three of these are NOAA satellites (NOAA-J, NOAA-K, and NOAA-L) that comprise the NOAA Polar Operational Environmental System (POES). These NOAA satellites all carry the Advanced Very High Resolution Radiometer (AVHRR) sensor that is used for measuring vegetation densities, crop yields, ocean temperatures, forest fire danger zones, and snow cover.

The other seven are NASA satellites: Landsat 5, Landsat 7, Terra, Tropical Rainfall Measuring Mission (TRMM), Earth Probe—TOMS, Quick Scatterometer (QuikScat) and Earth Observing-1 (EO-1). Landsat 7 is used for general Earth observations including forestry, crop monitoring, land cover, land use, and watersheds.

It carries the Enhanced Thematic Mapper Plus (ETM+), which boasts improved data collection capabilities from previous Landsat missions. Terra differs from Landsat in that it is dedicated to observing process more than land features. Terra carries five different sensors, each having unique applications, ranging from land temperature and snow/glacier cover measurements (ASTER), to cloud cover and radiant energy (CERES), to pollution measurements (MOPITT), to aerosol and smoke plume imaging (MISR), to ocean productivity and temperature ranges (MODIS). The five sensors carried on the TRMM satellite are all committed to record tropical and subtropical atmospheric parameters such as rainfall, lightning, and cloud cover. The TOMS sensor carried on the Earth Probe craft observes rates of ozone depletion, daily UV exposure, UV-absorbing aerosols and data on dust, smoke, and ash in the troposphere. The SeaWinds sensor carried on QuikScat uses specialized radar to measure near-surface wind speed and direction.

International Remote Sensing Programs

International efforts have pioneered the development of active remote sensing satellites. Canadian Space Agency's RADARSAT-1 and the European Space Agency's Remote Sensing satellites (ERS-1 and-2) carry radar sensors that emit and record microwave signals, permitting observations independent of weather or daylight conditions.

France, India, Russia, Japan, and the China-Brazil team all operate successful passive satellite programs. France controls the Systeme Pour l'Observation de la Terre (SPOT), which is comprised of satellites SPOT 1, SPOT 2, and SPOT 4. The payload of the SPOT satellites consists of two high-resolution-visible (HRVIR) sensors that can operate in either panchromatic (SPOT pan) or multispectral (SPOT xs) modes with a resolution of 10-20 meters depending on the mode. SPOT has many applications, including land use, water resources research, coastal monitoring, crop production, and deforestation. SPOT-4 also carries a wide-angle (2000 km) system referred to as VEGETATION that will be used for international crop monitoring. The Indian Space Research Organization (ISRO) currently operates four Earth-observing satellites; the most recently launched (IRS-P4/Ocenasat) focuses on oceanic research. Other Earth-observing systems (EOS) include Russia's Resurs-O1 series, Japan's ADEOS system, and the CBERS satellite that is operated jointly by China and Brazil.

Commercial Satellite Systems

The U.S. government has encouraged the development of independent commercial satellites and many U.S. companies have designed and launched their own satellites. Orbital Imaging Corporation (ORBIMAGE) and Space Imaging, Inc. both have successful satellites in orbit that carry high-resolution sensors. ORBIMAGE operates two satellites. The first, OrbView-1, is designed to monitor atmosphere. The second, OrbView-2 (SeaStar), carries a sensor called SeaWiFS (Sea-viewing Wide Field-of-view Sensor) that was developed in conjunction with NASA. SeaWiFS is designed to monitor ocean temperature and productivity. Space Imaging operates one satellite, IKONOS, which boasts 1-meter resolution capabilities in the panchromatic (black and white) range and 4-meter resolution in the multispectral range. IKONOS has applications ranging from imaging coral reefs to aiding highway planning.

Remote Sensing Applications

The potential applications of these satellite sensors are vast. This section briefly describes some of the possible environmental applications, focusing on environmental enforcement, land use planning, forestry, agriculture, water resources, fisheries, wetlands, watersheds, climate change, and disaster management.

Environmental Enforcement

The U.S. Environmental Protection Agency (EPA) conducts four types of satellite and aerial remote sensing projects to support the Comprehensive Environmental Response, Compensation, and Liability Act (CERCLA, also known as the Superfund Act), the Resource Conservation and Recovery Act (RCRA), and in other EPA regulatory programs and investigations. The projects are: (1) emergency response to hazardous material release that requires rapid site assessment; (2) single-date analysis to update old data on the current conditions of the site; (3) intensive site analysis of current and historic images, to obtain an understanding of changing conditions over time; and (4) waste site inventories over large areas to locate possible disposal sites. Images from these projects can standalone or be used in conjunction with topographic maps, digital elevation data, and other features stored in GIS databases.

Further use of remote sensing (both satellite and aerial photography) as a tool in environmental forensics is discussed in a two-part paper by Brilis, et al. The paper outlines the general approach to be followed when planning the use of remote sensing in environmental forensics. The accuracy of locational data and the use of metadata are identified as two critical items to ensure that a final image can withstand veracity issues when used for courtroom presentation.

Recently, interest has developed in using satellites to monitor and enforce multilateral environmental agreements (MEAs), such as the Kyoto Protocol. Remote sensing data may be used in the future to ensure compliance with MEA requirements by both direct enforcement and by more indirect means, such as deterring non compliance through high levels of transparency.

Land Use Planning and Change

Passive sensors, including those on the NOAA-AVHRR, IKONOS, Landsat, and SPOT satellites, are used in a broad range of forest and land use applications. These applications include estimations of primary production, biomass, crop yields, and to chart vegetation type, deforestation, desertification, forest boundaries, forest harvest, soil erosion, and bush or forest fires. Landsat 7's EMT+ sensor is especially useful in studying land use change because its data has been archived since the first Landsat mission in 1972. Passive sensors have also been used to observe and monitor changes associated with storm, flood, and fire damage.

Forestry

Forestry applications for passive remote sensors include tree species surveys, monitoring clear cut operations, planning and observing burn areas, and studying successional forest growth. The U.S. Forest Service (USFS) relies primarily on the data from Landsats 5 & 7 for forest monitoring because of the low cost and large scene size. Landsat data is particularly applicable to forest change monitoring because data from previous Landsat missions is archived and available for accurate comparison with data from the current Landsat mission. The USFS also uses SPOT data in conjunction with Landsat data to increase the level of detail in sensitive areas. Active sensors, carried on the RADARSAT and ERS satellites, are capable of making course scale distinctions between cover types such as late successional forests, newly planted forests, clear cut forests, burn areas, agricultural areas, and deserts. Active sensors are valuable tools for monitoring crop regulation compliance, forest clearing, and for taking general inventories of world forest densities.

Agriculture

The United States Department of Agriculture (USDA) is conducting research to determine the potential uses of remote sensing (both aerial and satellite) in the agricultural sector. Promising applications include measuring leaf area indices (LAI—a quantitative indicator of leaf stress), identifying soil properties by their spectral signals, evaluating crop productivity, and providing a valuable data source for crop simulation models. A high-tech type of farming known as "precision agriculture," uses satellite data to characterize specific sections of a field by certain variables (such as water or nutrient levels). Once the characteristics and geographic coordinates of the field section are in a computer, additions such as water, pesticides, and fertilizers can be efficiently controlled in response to the specific needs of each section thereby reducing the amount of pollutants introduced to the environment while producing healthier crops.

Water Resources and Fisheries

SeaWiFS is designed to monitor oceans and track water indicators such as turbidity, sediment load and transport, primary production by marine phytoplankton, algal blooms, chlorophyll content, dissolved oxygen, and pH. Other applications include managing coral reefs, monitoring pollution and oil spills, and characterizing and monitoring short-term and long-term fish habitat. Terra's MODIS and AVHRR sensors record observations of sea surface temperature, which is directly relevant to fisheries due to individual species' temperature requirements for survival and propagation. The sensor may also help predict migration routes. Active sensing technologies are capable of measuring sea level, wave height, surface wind speed, current fronts, eddies, and surface temperature, as well as locating ocean floor features such as trenches and seamounts. Active sensors have also been used to track oil spills, effluent discharges, and algal blooms.

Wetlands and Watersheds

Wetlands monitoring may employ a combination of land-observation and ocean-observation satellites. ETM+ data can be used to delineate wetland areas, make topographical observations, and to detect illegal development. Active systems can provide consistent and accurate observations of dynamic wetland parameters such as tidal and seasonal patterns, climate, hydrology, topography, vegetation, and soil type. Satellite data and images can also be used to delineate the flow of water through watersheds, and can even be used to track pollutants. Furthermore, using algal productivity as an indicator, scientists are able to monitor whether high levels of nutrients pollute areas of a watershed.

Climate Change

In the past decade, various ozone-monitoring sensors have been launched to study global climate cycles. These include the TOMS sensor and many of the sensors on Terra and future EOS satellites. AVHRR data from NOAA's POES satellites is used in conjunction with RADARSAT to monitor the polar ice sheets and iceberg movements. The EOS satellites, beginning with the Terra, were designed specifically for monitoring climate conditions, including the observation of aerosols, cloud cover, fires, ocean productivity, pollution, solar radiation, sea ice, and snow cover.

Disaster Management and Emergency Response

Remote sensing technologies can provide the government with the ability to avoid much of the damage caused by unforeseen natural disasters. While weather satellites have monitored hurricanes and tornados since the 1960s, other satellite sensors, such as ETM+ and MODIS, have potential applications for disaster management and response. Scientists have used ETM+ data to monitor patterns in floods, droughts, beach erosion, and volcanic activity over time. MODIS and ASTER data can forecast severe weather with a great degree of reliability, potentially saving states millions of dollars in unnecessary evacuation and emergency response. For forest fire emergencies, TOMS data can identify and monitor the occurrence of forest fires, especially in remote areas, while AVHRR data can create maps denoting fire-susceptible areas. NOAA-POES and NOAA-GOES (Geostationary Operational Environmental Satellite) are used to make weather observations including predicting local weather, tracking weather in real time globally and locally, understanding and predicting hurricanes and other severe weather, studying phenomena such as El Nino, La Nina, the Gulf Stream and other global current patterns, and observing the dynamics between the land temperature, ocean processes, and the atmosphere.

SATELLITE ORBIT PROPAGATION

Modeling satellite orbit propagation can be accomplished tram a deterministic model or a statistical model of numerous orbital ephemerae. Because the deterministic approach is so well known in classical physics, we do not seek to repeat the equations and algorithms. While fundamental in nature, these equations would still take some ten pages or so for presentation and more for explanation. The deterministic approach is well grounded but complex (Hoots and Roehrich 1980). One of the earliest works on deterministic modeling of satellite orbits around the Earth was by Brouwer (1959). A brief history of the fundamentals of Brouwer's research and the subsequent modifications by others is presented by Hoots (1981). Tapley et al. (2004) provide a thorough discussion of statistical approaches for modeling satellite orbit propagation. Notable works by others (e.g., Emery et al. 1989) for specific sensors on remote sensing satellites have also been conducted. Those interested in developing their own satellite propagation or sensor models are encouraged to consult these fundamental works.

The orbital propagation solution and sensor viewing geometry developed for the Remote Sensing Hazard Guidance System (Hodgson and Kar 2008) was used in this research. We have built our model based on these works but modified it for efficiency and accuracy. The age of the satellite ephemerae (i.e., the location and trajectory of a satellite observed at one moment) is one of the major factors influencing satellite orbit propagation accuracy. Predictions of near-future satellite locations (e.g., hours later) are trivial and very accurate, while predictions for months ahead require greater complexity and numerous additional factors. Our model was based on other demanding applications but was used in a very ideal context here. For this research, we used orbital ephemerae that were collected nightly and thus are less than 24 hours old. Our orbital predictions for day 'n' used the ephemerae from day 'n'-1 (i.e., the preceding day). By using daily ephemerae for predictions within the next 24 hours at the most, we also obviate the major sources of error in satellite orbit prediction, such as satellite maneuvers.

The first step in the orbital propagation process is to obtain the orbits of all the satellites of interest. The orbital ephemerides are obtained on a nightly basis for all remote sensing satellites. The satellite orbits are converted into a Cartesian Earth centered state vector at the particular epoch of interest. The Cartesian state vectors are used to compute classical orbit elements for each of the satellites. These classical orbit elements may be used to propagate the satellite trajectory forward or backward in time from the epoch specified. In addition, the time and state vector for a particular satellite may be determined using angles from the satellite location at the epoch of interest. For example, the state of a satellite when it reaches its northern most position may be determined from the angle measured from the initial position to the point in the orbit where the true anomaly plus argument of periapsis is 90 degrees. Using these classical orbit elements, the parts of the orbit where the spacecraft is on the dark side of the Earth or too far north to be imaged may be quickly eliminated.

The search algorithm then finds the point on the orbit where the satellite is closest to the landfall point defined on the surface. At this point, the range rate is zero and transitioning from negative to positive. These geometrical parameters may be easily computed from the landfall point location and state vectors of the satellite obtained from orbit element propagation. Various geometrical parameters, such as sun elevation angle or range, may be examined to determine if they satisfy the constraints specified. If they are out of tolerance, the satellite will not have an opportunity to collect imagery at this location.

Once the minimum-range point is found, as determined from conic orbit elements, a new initial epoch is computed. The ephemerae database is interrogated for a new satellite state vector for the date/time calculated using the conic orbit element. The entire search algorithm is then repeated using precision satellite state vectors at the exact time of overflight. This insures that the most accurate results are obtained, and conic orbit propagation is only used for relatively short distances.

SATELLITE VIEWING OPPORTUNITIES

To empirically determine the likelihood of future collection opportunities, the spatial and temporal distribution of historic landfall points were assumed to occur in the year 2005. The year 2005 is arbitrary but was convenient for this research as we had a complete daily collection of the satellite orbital ephemerae needed to predict the orbital tracks of each satellite and, subsequently, the satellite viewing opportunities. As noted earlier, using daily orbital ephemerae allows for very high spatial/temporal accuracy. The use of any other year may change the collection opportunities for a single hurricane event but not for the set of landfall events, unless the orbits of the satellites were temporally correlated.

For each day and geographic landfall location of a hurricane, the viewing opportunity for each of three satellites (i.e., Ikonos-2, Quickbird-2, and Orbview-3) was modelled. Determining if the geographic location can be imaged is solved by first propagating the satellite vehicle orbit during the days following the hurricane landfall, and modeling the viewing geometry for the sensors onboard the satellite vehicle. Most satellites with high spatial resolution sensors have a similar orbital period (90-100 min) resulting in similar repeat intervals on the ground (satellite image providers routinely advertise a 2-3 day repeat visit).

Imaging opportunities vary among satellite sensors with variations in altitude and off-nadir viewing. The OSA sensor on Ikonos-2 and Orbview-3 can point 45-degrees off-nadir as compared to the 28-degrees of the BGIS200 on Quickbird-2. For most mapping applications (except for stereoviewing applications), off-nadir viewing angles of less than 26 degrees are desirable. At greater angles, the ground surface obscured by the sides of land cover features (buildings, vegetation, etc.) is problematic. The potential swath coverage depicts the possible image collection opportunities; however, it does not represent the area-field-of-view, also called swath width, which would be imaged by most high spatial resolution sensors. The swath width is always smaller (e.g., 16.5-km in width for Quickbird-2 at nadir) than the potential swath coverage.

The likelihood of satellite imagery collection was examined under several scenarios. First, we assumed a single image provider, and a maximum viewing angle of 26 degrees off-nadir was permitted. Historically, remote sensors in a hurricane disaster context have sought imagery from near-nadir views rather than oblique imagery. The 26-degree criterion has been used informally in emergency operations centres. Second, any of the three satellites were allowed but still imposing a 26-degree off-nadir limit. Finally, collection opportunities from 1) single satellites and 9) all three satellites were examined with the maximum viewing angle allowed by the sensors.

Restricted Viewing Angles

If an emergency response agency relied on a single satellite provider to collect imagery over the hurricane landfall area, the likelihood of obtaining imagery within the first 24 hours ranges from 16 to 18 percent. The likelihood increases to 37 to 42 percent during the first 48 hours and from 53 to 67 percent in the first three days. Even with the restriction of 26 degrees maximum viewing angle, the Ikonos-2 satellite-sensor combination has a higher likelihood of collection as its altitude is higher (i.e., 679 km versus-450 km for the other two).

A higher altitude allows the sensor to point at farther planimetric distances with the same viewing angle. Relying on one satellite-sensor combination with this 26 degree off-nadir viewing limit results in an image collection during the first three days of, at best, only 2 out of 3 (-67 percent) hurricane disasters. However, if satellite imagery from any of the three satellites were permitted, then the likelihood increases to 38 percent, 74 percent, and 91 percent within the first 24, 48, and 72 hours after hurricane landfall.

Maximum Viewing Angles

Not surprisingly, when more extreme viewing angles are allowed (e.g., when the agency must have some imagery regardless of whether portions of the landscape are obscured), the likelihood of collecting imagery increases. Ikonos-2 and Orbview-3 can point off-nadir at angles of 45 degrees. The likelihood of collection in the first 24 hours increases to 39 percent using either of theses satellites. The likelihood increases to 97 percent for either Ikonos-2 or Orbview-3 within the first 72 hours after landfall. By relying on all three satellite image providers, the likelihood of collecting in the first 24 hours is 61 percent.

This research estimated the likelihood of using one or more (up to three) satellite imagery sensors for collecting post-hurricane event imagery in the first 24, 48, and 72 hours of landfall. Relying on a single satellite provider would result in only a 39 percent chance of collecting in the first 24 hours, if using maximum off-nadir viewing angles. To improve the collection likelihoods for post-event imagery, multiple satellite providers must be relied upon. In the concept of operations (CONOPS) model, this would inevitably require pre-existing contracts (pre-existing to enable a rapid agreement to collect and pay for the collection) with multiple satellite image providers. Since the completion of this research, Orbview-3 satellite imaging sensor has malfunctioned and is considered inoperable. However, GeoEye-1, with a.41 cm spatial resolution and a 681-km altitude and off-nadir viewing angle of up to 61 degrees is operational. Using the GeoEye-1 satellite in lieu of Orbview-3 would result in likelihood estimates slightly better than Ikonos-2. Other high spatial resolution satellite-sensors, such as Worldview-1, have recently been launched, providing viewing opportunities from at least four different U.S. commercially controlled satellites.

A logical planning scenario might be to use multiple satellite-sensor contracts and stratify the potential damage area into image collection priorities. An image collection with a near-nadir (low off-nadir angles) vantage might be used for the high-priority areas, while greater off-nadir angle opportunities are used for low priority areas.2 Focusing on large-impact areas would further increase the likelihood of collecting imagery across a disaster area.

A final probing scenario for response agencies might be to use both aerial and satellite imaging platforms in a well defined and coordinated approach. By quickly modeling (e.g., GIS-based without imagery the likely damage area, the collection area could be spatially defined, stratified, and prioritized. Available airborne and satellite-sensors could be utilized in a stratified (both temporally and spatially) approach to collect imagery as part of the post-event response.

There is a revolution underway in Caddo archaeology, and its name is remote sensing. The use of remote sensing technologies on Caddo archaeological sites-in conjunction with archaeological excavations-is producing, and will continue to produce, unprecedented characterizations of the internal spatial structure and organization of Caddo villages and mound centres. We illustrate the importance of remote sensing for the present and future of Caddo archaeology by focusing on the remote sensing and archaeological findings from one Caddo village on the Red River in Bowie County, Texas. The Hill Farm site is part of the larger Hatchel site village and was occupied by the Nasoni Caddo between ca. A.D. 1600 and 1700.

The Nasoni Caddo village was visited by Europeans in 1687, 1690, and 1691, most famously by the Spanish expedition led by Don Domingo Teran de los Rios. On this expedition (Hatcher 1932, 1999), a detailed map of this village was drawn that showed a temple mound and a number of village compounds (with house structures, granaries, and outdoor ramadas or arbors). A convergence of large-scale remote sensing (Grealy and Conyers 2006), the smaller-scale remote sensing, geoarchaeological findings (Guccione and Hays 2006), and the archaeological discoveries at the Hill Farm site and other places at the Hatchel site (Perttula 2005) have led us to the specific identification of the Hill Farm site with two of the village compounds depicted on the 1691 Teran map. This in turn opens the door to the possibility that continued archaeological and remote investigations can, in effect, locate and ground truth the 1691 Teran map and different parts of the Nasoni Caddo village.

Remote Sensing in Caddo Archaeology

Archaeogeophysical prospecting using magnetometry, electromagnetic conductivity, electrical resistivity, ground-penetrating radar, and magnetic susceptibility have all been shown to be useful geophysical techniques for locating buried architectural remains on prehistoric Caddo mound sites, such as the George C. Davis site and Horace Cabe Mounds (Walker and Perttula 2007) in northeastern Texas, and at various sites on the Grandview project in southwestern Arkansas. These techniques measure certain physical properties of the soil, such as magnetic conditions, electromagnetic conductivity, or electrical resistivity. Given the proper soil conditions, archaeological features such as fire hearths, post holes, and storage pits differ from the surrounding soil matrix, and can be recognized by distinctive spatial signatures in geophysical data sets from archaeological sites.

At the George C. Davis site, geophysical surveys, using a portable cesium magnetometer, on large blocks of land have proved successful in locating both large and small architectural features, particularly numerous Caddo structures. This work is contributing new information on the overall layout of structures around the mounds and across this large prehistoric site. In some settings, the geophysical survey has successfully located archaeological features in areas that previous subsurface excavations failed to locate (Walker et al. 2003). Shaded relief maps of the magnetic data have also identified a number of unique architectural forms at the site, such as the so-called Button Houses (Creel et al.

2005), that can be compared to the architectural plans of the more than 50 previously excavated structures at the George C. Davis site.

These geophysical examples of Caddo structures are 10-15 m diameter circular structures with four internal magnetic returns that are spatially patterned on the interior of the structure to encircle a central hearth; these no doubt represent large post hole features (support posts) or interior pits. The magnetometer data from the Mound B area at George C. Davis are providing new information about the construction phases of the mound. These data have detected a trench of some kind or a small mound of earth that apparently served to anchor the mound to a specific place on the ground. The remote sensing work has also identified a second ramp on the southern flanks of the mound (Schultz et al. 2004).

On the Grandview project, the Arkansas Archeological Survey used a variety of geophysical surveying equipment to measure magnetic susceptibility, gradiometry, electrical resistance, and electromagnetic conductivity of key areas at Caddo mound and nonmound settlements, and achieved impressive results. The geophysical surveying has detected the locations of burned Caddo houses, fireplaces or hearths within the houses, burial pits, nonburial pits of various sizes and shapes, stockade lines around the Tom Jones mound (3HE40) (Lockhart 2007), and possible paths across the sites.

Ground-truthing excavations of several of the subsurface anomalies identified in the Grandview geophysical survey have been successful in identifying archaeological features, including burned houses with "heaps of fired daub up to a meter deep in places, pieces of carbonized logs, and a floor laden with ash and domestic debris" (Schambach 2001:10). Other excavations exposed clay-lined hearths, and lines of post holes, while excavations placed over "reasonably clear images [from the remote sensing] of house remains netted... the archaeological remains of four houses" (Schambach 2001:10).

The Archaeology of the Hill Farm Site within the Larger Hatchel Site Village

Before discussing the remote sensing findings and our interpretation of them, we review what is known about the archaeology of the Hill Farm site. The site is only one part of the much larger Hatchel site village occupied by the Nasoni Caddo from the eleventh to the eighteenth century (Perttula 2005). The village was apparently comprised of a number of farmstead compounds spread out over many acres, with domestic structures, granaries, arbors, and ramadas. The farmstead compounds may have been separated by compound fences or dividers of trees and bushes. The Hill Farm site has two areas, Areas A and B, with substantial archaeological deposits in both.

The first archaeological investigations at the Hill Farm part of the Hatchel village was completed by the Works Progress Administration (WPA) and the University of Texas at Austin in 1939 (TARL n.d.a). The work was on a small Caddo mound, 21 m in diameter and 0.9 m high, which is 915 m southeast of the main platform mound at the Hatchel site and 610 m southeast of WPA excavations in Village Plot 1 (Perttula 2005). The mound (Mound 2) is in Area A at the Hill Farm site; its exact location is currently unknown. Three other small mounds (0.6-0.9 m in height) were nearby. Just east of Mound 2 was a borrow pit of unspecified size (TARL n.d.a).

The WPA excavations in Mound 2 were in two blocks at the northern and southern parts of the mound. The northern strip covered 150 ft^2 and the southern excavation was 900 ft^2 in size. These excavations disclosed that the mound was constructed in one episode of a brown clay fill that was 0.9 m thick. It was probably constructed to bury a structure or important public building, although the details of this building are not known. The mound fill was placed atop a 25 cm thick habitation and midden deposit of Late Caddo Texarkana phase age (ca. A.D. 1400-1700). No structural features or post holes were recorded in these submound deposits.

A single extended burial (Burial 1) was excavated in the southern Mound 2 strip. This adult burial originated at the top of the habitation/midden deposits-prior to mound construction-and the grave pit reached to 1.37 m below surface (bs). No funerary offerings were placed with this Caddo individual. There were seven sherds, including a vessel lug, in the burial pit fill. Only 324 artifacts were recovered in the Mound 2 WPA excavations (none of the archaeological deposits were screened during the WPA work). This included 212 ceramic vessel sherds, one celt, 25 pieces of lithic debris, 71 animal bones, and 15 pieces of freshwater mussel shell.

In the 1950s, according to records on file at the Texas Archeological Research Laboratory, University of Texas, two avocational archaeologists (R. King Harris, from Dallas, Texas, and Pete Miroir, from Texarkana, Texas) conducted excavations in what appears to have been Area B at the Hill Farm site. They apparently encountered several Caddo burials and very large post holes from a large structure. No more details on this work are available, unfortunately.

More recently, Gilmore and McCormick reported that an unknown number of Caddo burials had been looted at the Hill Farm site in the 1970s. Their location within the site is also unknown. In 2005, renewed archaeological investigations at the Hill Farm site were completed as part of the Bowie County Levee Realignment project sponsored by the Tulsa District of the U.S. Army Corps of Engineers. During this work (Sundermeyer et al. 2006), 78 shovel tests were excavated in Area A (27 of which contained prehistoric archaeological materials), with another 127 shovel tests (51 contained prehistoric archaeological remains) in Area B.

Additionally, a magnetometer survey was completed in both Area A (6,000 m^2, ca. 1.5 acres) and Area B (3,600 m^2, 0.9 acres) by Walker and Schultz (2006). It is the findings from both the shovel testing and the remote sensing that we discuss in the remainder of this chapter. The Hill Farm site contains exceptionally well-preserved and buried Late Caddo Texarkana phase archaeological deposits. The archaeological materials are incorporated in Red River alluvial sediments of the abandoned and recently defined Cemetery Meander (Guccione and Hays 2006). These deposits apparently began to accumulate sometime after ca. A.D. 1240.

The site lies on sandy loam point bar alluvial deposits adjacent to an ancient abandoned channel of the Red River. The archaeological deposits are marked by numerous structures of varying sizes, pit features, midden deposits, at least one constructed earthen mound (Mound 2 at Hatchel) and as many as three other small mounds, a borrow pit, and one extended burial.

The recent shovel testing reported by Sundermeyer et al. (2006) identified two distinct areas (A and B) at the Hill Farm site. Area B, the northern area, covers approximately 13,000 m^2 (3.2 acres), while the southern part of the site (Area A) is about 11,000 m^2 (2.7 acres) in size. The two different site areas are about 160 m apart. Archaeological materials occur to depths of 80100 cm bs. The vast majority of these remains are buried, coming primarily from 20-80 cm bs. There are two concentrations of ceramic vessel sherds in Area A, with one large and heavy-density cluster covering ca. 2,400 m^2 in the centre of this village locality. A much smaller and lighter scatter of sherds is ca. 15-30 m south of the principal sherd concentration. In Area B, there are 4-5 distinct concentrations of ceramic vessel sherds. These concentrations are 15-40 m apart from each other within this village area, and most of them are situated in the southern part of Area B.

SATELLITE REMOTE SENSING

In the 1960s, a revolution in remote sensing technology began with the deployment of space satellites.

From their high vantage-point, satellites have a greatly extended view of the Earth's surface. The first meteorological satellite, TIROS-1, was launched by the United States using an Atlas rocket on April 1, 1960. This early weather satellite used vidicon cameras to scan wide areas of the Earth's surface. Early satellite remote sensors did not use conventional film to produce their images. Instead, the sensors digitally capture the images using a device similar to a television camera. Once captured, this data is then transmitted electronically to receiving stations found on the Earth's surface. Today, the GOES (Geostationary Operational Environmental Satellite) system of satellites provides most of the remotely sensed weather information for North America. To cover the complete continent and adjacent oceans two satellites are employed in a geostationary orbit.

The western half of North America and the eastern Pacific Ocean is monitored by GOES-10, which is directly above the equator and 135° West longitude. The eastern half of North America and the western Atlantic are cover by GOES-8. The GOES-8 satellite is located overhead of the equator and 75° West longitude.

Advanced sensors aboard the GOES satellite produce a continuous data stream so images can be viewed at any instance. The imaging sensor produces visible and infrared images of the Earth's terrestrial surface and oceans. Infrared images can depict weather conditions even during the night. Another sensor aboard the satellite can determine vertical temperature profiles, vertical moisture profiles, total precipitable water, and atmospheric stability.

In the 1970s, the second revolution in remote sensing technology began with the deployment of the Landsat satellites. Since this 1972, several generations of Landsat satellites with their Multispectral Scanners (MSS) have been providing continuous coverage of the Earth for almost 30 years.

Current, Landsat satellites orbit the Earth's surface at an altitude of approximately 700 kilometres. Spatial resolution of objects on the ground surface is 79 x 56 meters. Complete coverage of the globe requires 233 orbits and occurs every 16 days.

The Multispectral Scanner records a zone of the Earth's surface that is 185 kilometres wide in four wavelength bands: band 4 at 0.5 to 0.6 micrometers, band 5 at 0.6 to 0.7 micrometers, band 6 at 0.7 to 0.8 micrometers, and band 7 at 0.8 to 1.1 micrometers. Bands 4 and 5 receive the green and red wavelengths in the visible light range of the electromagnetic spectrum.

The last two bands image near-infrared wavelengths. A second sensing system was added to Landsat satellites launched after 1982. This imaging system, known as the Thematic Mapper, records seven wavelength bands from the visible to far-infrared portions of the electromagnetic spectrum. In addition, the ground resolution of this sensor was enhanced to 30 x 20 meters. This modification allows for greatly improved clarity of imaged objects.

The usefulness of satellites for remote sensing has resulted in several other organizations launching their own devices. In France, the SPOT (*Satellite Pour l'Observation de la Terre*) satellite program has launched five satellites since 1986. Since 1986, SPOT satellites have produced more than 10 million images.

SPOT satellites use two different sensing systems to image the planet. One sensing system produces black and white panchromatic images from the visible band (0.51 to 0.73 micrometers) with a ground resolution of 10 x 10 meters. The other sensing device is multispectral capturing green, red, and reflected infrared bands at 20 x 20 meters. SPOT-5, which was launched in 2002, is much improved from the first four versions of SPOT satellites. SPOT-5 has a maximum ground resolution of 2.5 x 2.5 meters in both panchromatic mode and multispectral operation.

Radarsat-1 was launched by the Canadian Space Agency in November, 1995. As a remote sensing device, Radarsat is quite different from the Landsat and SPOT satellites. Radarsat is an active remote sensing system that transmits and receives microwave radiation. Landsat and SPOT sensors passively measure reflected radiation at wavelengths roughly equivalent to those detected by our eyes.

Radarsat's microwave energy penetrates clouds, rain, dust, or haze and produces images regardless of the Sun's illumination allowing it to image in darkness. Radarsat images have a resolution between 8 to 100 meters. This sensor has found important applications in crop monitoring, defence surveillance, disaster monitoring, geologic resource mapping, sea-ice mapping and monitoring, oil slick detection, and digital elevation modelling.

PRINCIPLES OF OBJECT IDENTIFICATION

Most people have no problem identifying objects from photographs taken from an oblique angle. Such views are natural to the human eye and are part of our everyday experience. However, most remotely sensed images are taken from an overhead or vertical perspective and from distances quite removed from ground level. Both of these circumstances make the interpretation of natural and human-made objects somewhat difficult. In addition, images obtained from devices that receive and capture electromagnetic wavelengths outside human vision can present views that are quite unfamiliar.

To overcome the potential difficulties involved in image recognition, professional image interpreters use a number of characteristics to help them identify remotely sensed objects. Some of these characteristics include:

Shape: this characteristic alone may serve to identify many objects. Examples include the long linear lines of highways, the intersecting runways of an airfield, the perfectly rectangular shape of buildings, or the recognizable shape of an outdoor baseball diamond.

Size: noting the relative and absolute sizes of objects is important in their identification. The scale of the image determines the absolute size of an object. As a result, it is very important to recognize the scale of the image to be analysed.

Image Tone or *Colour*: all objects reflect or emit specific signatures of electromagnetic radiation. In most cases, related types of objects emit or reflect similar wavelengths of radiation. Also, the types of recording device and recording media produce images that are reflective of their sensitivity to particular range of radiation. As a result, the interpreter must be aware of how the object being viewed will appear on the image examined. For example, on colour infrared images vegetation has a colour that ranges from pink to red rather than the usual tones of green.

Pattern: many objects arrange themselves in typical patterns. This is especially true of human-made phenomena. For example, orchards have a systematic arrangement imposed by a farmer, while natural vegetation usually has a random or chaotic pattern.

Shadow: shadows can sometimes be used to get a different view of an object. For example, an overhead photograph of a towering smokestack or a radio transmission tower normally presents an identification problem.

This difficulty can be over come by photographing these objects at Sun angles that cast shadows. These shadows then display the shape of the object on the ground. Shadows can also be a problem to interpreters because they often conceal things found on the Earth's surface.

Texture: imaged objects display some degree of coarseness or smoothness. This characteristic can sometimes be useful in object interpretation. For example, we would normally expect to see textural differences when comparing an area of grass with a field corn. Texture, just like object size, is directly related to the scale of the image.

DATA FORMATS FOR DIGITAL SATELLITE IMAGERY

Digital data from the various satellite systems supplied to the user in the form of computer readable tapes or CD-ROM. As no worldwide standard for the storage and transfer of remotely sensed data has been agreed upon, though the CEOS (Committee on Earth Observation Satellites) format is becoming accepted as the standard. Digital remote sensing data are often organised using one of the three common formats used to organise image data.

For an instance an image consisting of four spectral channels, which can be visualised as four superimposed images, with corresponding pixels in one band registering exactly to those in the other bands. These common formats are:

- Band Interleaved by Pixel (BIP)
- Band Interleaved by Line (BIL)
- Band Sequential (BQ).

Digital image analysis is usually conducted using Raster data structures - each image is treated as an array of values. It offers advantages for manipulation of pixel values by image processing system, as it is easy to find and locate pixels and their values. Disadvantages becomes apparent when one needs to represent the array of pixels as discrete patches or regions, where as Vector data structures uses polygonal patches and their boundaries as fundamental units for analysis and manipulation. Though vector format is not appropriate to for digital analysis of remotely sensed data.

APPLICATIONS OF REMOTE SENSING DATA

- Conventional radar is mostly associated with aerial traffic control, early warning, and certain large scale meteorological data. Doppler radar is used by local law enforcements' monitoring of speed limits and in enhanced meteorological collection such as wind speed and direction within weather systems. Other types of active collection includes plasmas in the ionosphere. Interferometric synthetic aperture radar is used to produce precise digital elevation models of large scale terrain.
- Laser and radar altimeters on satellites have provided a wide range of data. By measuring the bulges of water caused by gravity, they map features on the seafloor to a resolution of a mile or so. By measuring the height and wave-length of ocean waves, the altimeters measure wind speeds and direction, and surface ocean currents and directions.
- Light detection and ranging (LIDAR) is well known in examples of weapon ranging, laser illuminated homing of projectiles. LIDAR is used to detect and measure the concentration of various chemicals in the atmosphere, while airborne LIDAR can be used to measure heights of objects and features on the ground more accurately than with radar technology. Vegetation remote sensing is a principal application of LIDAR.
- Radiometers and photometers are the most common instrument in use, collecting reflected and emitted radiation in a wide range of frequencies. The most common are visible and infrared sensors, followed by microwave, gamma ray and rarely, ultraviolet. They may also be used to detect the emission spectra of various chemicals, providing data on chemical concentrations in the atmosphere.
- Stereographic pairs of aerial photographs have often been used to make topographic maps by imagery and terrain analysts in trafficability and highway departments for potential routes.
- Simultaneous multi-spectral platforms such as Landsat have been in use since the 70's. These thematic mappers take images in multiple wavelengths of electro-magnetic radiation (multi-spectral) and are usually found on Earth observation satellites, including (for example) the Landsat programme or the IKONOS satellite. Maps of land cover

and land use from thematic mapping can be used to prospect for minerals, detect or monitor land usage, deforestation, and examine the health of indigenous plants and crops, including entire farming regions or forests.

- Hyperspectral imaging produces an image where each pixel has full spectral information with imaging narrow spectral bands over a contiguous spectral range. Hyperspectral imagers are used in various applications including mineralogy, biology, defence, and environmental measurements.
- Within the scope of the combat against desertification, remote sensing allows to follow-up and monitor risk areas in the long term, to determine desertification factors, to support decision-makers in defining relevant measures of environmental management, and to assess their impacts.

Geodetic

- Overhead geodetic collection was first used in aerial submarine detection and gravitational data used in military maps. This data revealed minute perturbations in the Earth's gravitational field (geodesy) that may be used to determine changes in the mass distribution of the Earth, which in turn may be used for geological studies.

Acoustic and Near-acoustic

- Sonar: *passive sonar*, listening for the sound made by another object (a vessel, a whale etc); *active sonar*, emitting pulses of sounds and listening for echoes, used for detecting, ranging and measurements of underwater objects and terrain.
- Seismograms taken at different locations can locate and measure earthquakes (after they occur) by comparing the relative intensity and precise timing.

To coordinate a series of large-scale observations, most sensing systems depend on the following: platform location, what time it is, and the rotation and orientation of the sensor. High-end instruments now often use positional information from satellite navigation systems.

The rotation and orientation is often provided within a degree or two with electronic compasses. Compasses can measure not just azimuth (i. e. degrees to magnetic north), but also altitude (degrees above the horizon), since the magnetic field curves into the Earth at different angles at different latitudes. More exact orientations require gyroscopic-aided orientation, periodically realigned by different methods including navigation from stars or known benchmarks.

Resolution impacts collection and is best explained with the following relationship: less resolution=less detail & larger coverage, More resolution=more detail, less coverage.

The skilled management of collection results in cost-effective collection and avoid situations such as the use of multiple high resolution data which tends to clog transmission and storage infrastructure.

Data Processing

Generally speaking, remote sensing works on the principle of the *inverse problem.* While the object or phenomenon of interest (the state) may not be directly measured, there exists some other variable that can be detected and measured (the observation), which may be related to the object of interest through the use of a data-derived computer model.

The common analogy given to describe this is trying to determine the type of animal from its footprints. For example, while it is impossible to directly measure temperatures in the upper atmosphere, it is possible to measure the spectral emissions from a known chemical species (such as carbon dioxide) in that region.

The frequency of the emission may then be related to the temperature in that region via various thermodynamic relations.

The quality of remote sensing data consists of its spatial, spectral, radiometric and temporal resolutions.

Spatial Resolution

The size of a pixel that is recorded in a raster image – typically pixels may correspond to square areas ranging in side length from 1 to 1,000 metres (3.3 to 3,300 ft).

Spectral Resolution

The wavelength width of the different frequency bands recorded – usually, this is related to the number of frequency bands recorded by the platform. Current Landsat collection is that of seven bands, including several in the infra-red spectrum, ranging from a spectral resolution of 0.07 to 2.1 ìm. The Hyperion sensor on Earth Observing-1 resolves 220 bands from 0.4 to 2.5 ìm, with a spectral resolution of 0.10 to 0.11 ìm per band.

Radiometric Resolution

The number of different intensities of radiation the sensor is able to distinguish. Typically, this ranges from 8 to 14 bits, corresponding to 256 levels of the gray scale and up to 16,384 intensities or "shades" of colour, in each band. It also depends on the instrument noise.

Temporal Resolution

The frequency of flyovers by the satellite or plane, and is only relevant in time-series studies or those requiring an averaged or mosaic image as in deforesting monitoring.

This was first used by the intelligence community where repeated coverage revealed changes in infrastructure, the deployment of units or the modification/ introduction of equipment. Cloud cover over a given area or object makes it necessary to repeat the collection of said location.

In order to create sensor-based maps, most remote sensing systems expect to extrapolate sensor data in relation to a reference point including distances between known points on the ground.

This depends on the type of sensor used. For example, in conventional photographs, distances are accurate in the centre of the image, with the distortion of measurements increasing the farther you get from the centre. Another factor is that of the platen against which the film is pressed can cause severe errors when photographs are used to measure ground distances. The step in which this problem is resolved is called georeferencing, and involves computer-aided matching up of points in the image (typically 30 or more points per image) which is extrapolated with the use of an established benchmark, "warping" the image to produce accurate spatial data. As of the early 1990s, most satellite images are sold fully georeferenced. In addition, images may need to be radiometrically and atmospherically corrected.

Radiometric Correction

Gives a scale to the pixel values, *e.g.* the monochromatic scale of 0 to 255 will be converted to actual radiance values. Atmospheric correction eliminates atmospheric haze by rescaling each frequency band so that its minimum value (usually realised in water bodies) corresponds to a pixel value of 0.

The digitizing of data also make possible to manipulate the data by changing gray-scale values.

Interpretation is the critical process of making sense of the data. The first application was that of aerial photographic collection which used the following process; spatial measurement through the use of a light table in both conventional single or stereographic coverage, added skills such as the use of photogrammetry, the use of photomosaics, repeat coverage, Making use of objects' known dimensions in order to detect modifications. Image Analysis is the recently developed automated computer-aided application which is in increasing use.

Object-Based Image Analysis (OBIA) is a sub-discipline of GIScience devoted to partitioning remote sensing (RS) imagery into meaningful image-objects, and assessing their characteristics through spatial, spectral and temporal scale. Old data from remote sensing is often valuable because it may provide the only long-term data for a large extent of geography. At the same time, the data is often complex to interpret, and bulky to store. Modern systems tend to store the data digitally, often with lossless compression. The difficulty with this approach is that the data is fragile, the format may be archaic, and the data may be easy to falsify.

One of the best systems for archiving data series is as computer-generated machine-readable ultrafiche, usually in typefonts such as OCR-B, or as digitized half-tone images. Ultrafiches survive well in standard libraries, with lifetimes of several centuries. They can be created, copied, filed and retrieved by automated systems. They are about as compact as archival magnetic media, and yet can be read by human beings with minimal, standardized equipment.

ACQUIRING REMOTE SENSING DATA

Remote sensors can be grouped according to the number of bands and the frequency range of those bands that the sensor can detect. Common categories of remote sensors include *panchromatic, multispectral, hyperspectral,* and *ultraspectral* sensors.

Panchromatic sensors cover a wide band of wavelengths in the visible light or near infrared light spectrum. An example of a single band sensor of this type would be a black and white photographic film camera.

Multispectral sensors cover two or more spectral bands simultaneously typically from 0.3 m to 14 m wide.

Hyperspectral sensors cover spectral bands narrower than multispectral sensors. Image data from several hundred bands are recorded at the same time offering much greater spectral resolution than a sensor covering wider bands.

Ultrasprectral sensors are still under development and not yet in use. These sensors will cover thousands of bands with an even narrower bandwidth than hyperspectral sensors.

Digital Image Data Delivery Systems

Electro-optical and spectral imaging scanners produce digital images with the use of detectors that measure the brightness of reflected electromagnetic energy. Scanners consist of one or more sensor detectors depending on type of sensor system used. One type of scanner is called a whiskbroom scanner also referred to as across-track scanners. It uses rotating mirrors to scan the landscape below from side to side perpendicular to the direction of the sensor platform, like a whiskbroom. The width of the sweep is referred to as the sensor swath.

The rotating mirrors redirect the reflected light to a point where a single or just a few sensor detectors are grouped together. Whiskbroom scanners with their moving mirrors tend to be large and complex to build. The moving mirrors create spatial distortions that must be corrected with preprocessing by the data provider before image data is delivered to the user. An advantage of whiskbroom scanners is that they have fewer sensor detectors to keep calibrated as compared to other types of sensors.

Another type of scanner, which does not use rotating mirrors, is the pushbroom scanner also referred to as an along-track scanner. The sensor detectors in a pushbroom scanner are lined up in a row called a linear array. Instead of sweeping from side to side as the sensor system moves forward, the one dimensional sensor array captures the entire scan line at once like a pushbroom would. Some recent scanners referred to as step stare scanners contain two-dimensional arrays in rows and columns for each band. Pushbroom scanners are lighter, smaller and less complex because of fewer moving parts than whiskbroom scanners. Also they have better radiometric and spatial resolution. A major disadvantage of pushbroom scanners is the calibration required for a large number of detectors that make up the sensor system.

Data Processing, Interpretation and Analysis

Much of the technical work of remote sensing involves pre-processing and applying radiometric and geometric corrections to imagery to compensate for errors due to factors such as atmospheric interference of incoming radiation and sensor and data stream irregularities. Once such corrections are applied, imagery must be *georeferenced*. Georeferencing is the process of taking the image in its raw format (rows and columns of data) and linking it to the land that it covers. Images are georeferenced by linking spatially distributed control points in the satellite image to points on base maps or points referenced in the field through global positioning systems. The raster data in the image is thereby registered to a Cartesian coordinate system, and can be combined with other georeferenced data sets in a geographic information system.

The processed data can now either be visually interpreted or classified using manual or automated processes.

The main elements of visual image interpretation involve gradients of tone or colour, resolution, size and shape, texture and pattern, site and association, and height and shadows. Given their knowledge of the characteristic spectral signatures of different land cover types, scientists my inspect black and white images of each band separately in order to identify features and patterns.

For many purposes, data that is collected from the earth's surface, which represents a continuous variation, needs to be categorized (de Sherbinin *et al.* 2002). Image classification is the process of creating discrete classes or categories of land cover, utilizing information from some or all of the bands to group together pixels with similar spectral signatures.

Supervised classification entails providing the software with sample pixels that represent specific features, such as boreal forest, and then having the computer classify every pixel with a similar *spectral signature* as boreal forest.

Analysts may also use images from different seasons in order to discriminate vegetation cover types that have different phenologies, such as deciduous and evergreen forests. In unsupervised classification, the analyst specifies the desired number of classes, and the computer automatically sorts the pixels according to their spectral signatures. The analyst then labels the resulting groups based on some local knowledge of the land cover patterns.

Once classified, it is necessary to verify or validate that the output product accurately represents the actual composition, content, structure or land surface characteristics being mapped. Validation requires either field visits, ground-truthing or comparing the classified image with existing maps or images of sufficient detail. Statistics can be derived for the classified imagery indicating the general and specific (class-wise) agreement between the pixels or classes used, letting the user know which were classified correctly and which ones were not.

Validation results are also sometimes presented as a percentage value associated with the map that communicates how accurate the map is on a per pixel basis. Since the highest confidence rankings reported by satellite land cover data sets are between 85% and 90% (for the easiest types of land cover to classify), for an image with a per pixel accuracy of 85% the probability that one pixel out of four is incorrectly classified is close to 0.50. The output of remote sensing data analysis can be presented in a variety of ways including a printout of the enhanced image itself a spatial database, summary statistics and/or graphs. The output data can be integrated with a geographic information system (GIS) database for further analysis.

SENSORS CHARACTERISTICS

A sensor is characterized by its spectral properties (number and placement of bands), its orbital altitude and path, its swath width, and its *spatial resolution*. Spatial resolution is measured in terms of the size of one pixel projected on the ground. Spatial resolution is directly tied to the size of the features that can be resolved (or "seen") on the ground. The higher the resolution, the less likely that there will be "mixed pixels" in which radiances effectively represent an average of land cover types in the ground area represented by that pixel (*e.g.*, half lake and half forest). Commercial high resolution sensors have a spatial resolution in the 0.6-10 meter range, medium resolution sensors fall in the 10-50 meter range, and low resolution sensors have greater than 50 meter resolution.

Until the advent of the commercial satellites IKONOS and QuickBird, with resolutions of one square meter or finer, high resolution imagery was the exclusive province of intelligence-gathering agencies. Most social science applications do not command the financial resources required to obtain such high resolution data, nor are images of this resolution generally required, except perhaps in the area of international relations, law and policy. Most social science research tends to utilize polar-orbiting satellites with medium spatial resolution, such as the Landsat, SPOT, TERRA and, more recently, AQUA satellites. They provide good spectral and ground resolution, with multiple visible, infrared, and panchromatic bands and pixel width ranging from 5 to 30 meters.

Bibliography

Albritton, C. C. : *The Fabric of Geology: Reading*, Mass., Addison-Wesley Publishing Co., 1963.

Allen J.: *Cultural Industries and the Production of Culture,* Taylor & Francis, 2007.

Bhagwat, S.B. : *Foundation of Geology,* Global Vision Pub, Delhi, 2009.

Bhatta, B : *Remote Sensing and GIS,* Oxford University Press, Delhi, 2008.

Bogli, Alfred: *Karst Hydrology and Physical Speleology*, Springer-Verlag, New York, 1980.

Curran, H. Allen, Young, Drew M., Justus, Philip S., and Garver, John B., Jr.: *Atlas of Landforms,*John Wiley & Sons, New York, 2001.

Darlene Elaine: *Global Perspectives in Environmental Adult Education*, New York, Peter Lang, 2003.

Davies R. J., Cartwright J., Stewart S. A, Underhill, J. R., Lappin M.:*3D Seismic Technology: Application to the Exploration of Sedimentary Basins,* Geological Society, London, Memoir, 2003.

Elachi, Charles : *Introduction to the Physics and Techniques of Remote Sensing*, New York, Wiley, 1987.

Elangovan K. : *GIS, Fundamentals, Applications and Implementations*, New India Publishing Agency, New Delhi 2006.

Evans, A.M.: *Ore Geology and Industrial Minerals, An Introduction.*, Blackwell Science, 1993.

Gontz, A.M.: *Evolution of Seabed Pockmarks in Penobscot Bay, Maine*, University of Maine, Orono, 2011.

Gurney, R. J. : *Remote Sensing in Hydrology*, New York, Van Nostrand Reinhold, 1991.

Gurugnanam, B. : *Essentials of Hydrogeology*, New India Pub, Delhi, 2009.

Hallam, A.: *A Revolution in the Earth Sciences*, New York, Oxford University Press, 1973.

Hansen D.M., Cartwright, J.A, and Thomas, D. 2004. The development of intersecting igneous sills from 3D seismic. Geological Society of London, Special Publication.*In:*

Harvey, Francis : *A Primer of GIS, Fundamental Geographic and Cartographic Concepts,* The Guilford Press, 2008.

Jakucs, Laszlo: *Morphogenetics of Karst Regions: Variants of Karst Evolution,* Halstead Press, John Wiley and Sons, New York, 1977.

Jamwal, Anil : *Principles of Remote Sensing,* Jnanada Prakashan, Delhi, 2010.

Jennings, Joseph N.: *Karst Geomorphology*, Basil Blackwell, Oxford and New York, 1985.

Kiefer, R. W. : *Remote Sensing and Image Interpretation,* New York, 1987.

Kondolf, G. Mathias; Hervé Piégay: *Tools in Fluvial Geomorphology*, Wiley, New York, 2003.

Kramer, Steven L.: *Geotechnical Earthquake Engineering*, Prentice-Hall, 1996.

Kuiper, Gerard P. : *The Earth as a Planet*, Chicago, Univ. of Chicago Press, 1954.

Kumar Yadav, Surendra : *Soil Ecology : Remote Sensing and Geographic Information System in Management*, APH, Delhi, 2007.

Leopold, L.B., M.G. Wolman, and J.P. Miller: *Fluvial processes in Geomorphology*, Freeman, San Francisco, CA, 1964.

Mack, Pamela Etter : *Viewing the Earth: The Social Construction of the Landsat Satellite System*, Cambridge, MIT Press, 1900.

McDonnell, R.A. : *Principles of Geographical Information Systems*, Oxford University Press, Oxford, 1998.

Milzow, C.: The Step-pool Morphology of a Steep Mountain Stream, Swiss Federal Institute of Technology, Zurich, Switzerland, 2004.

Nicholls, A.: *Low Attainers and the Teaching of Geography*, Sheffield, Geographical Association, 1988.

Nierenberg, W. A.: *Encyclopedia of Earth System Science*, San Diego, California: Academic Press, 1992.

Okoshi, T. : *Three-dimensional Imaging Techniques*, Academic Press Inc., New York, 1976.

Park, RG: *Foundations of Structural Geology*, Routledge, 2004.

Ramasamy, S. : *Remote Sensing in Water Resources*, Rawat, 2005.

Rondinelli, D.A.: *Applied Methods of Regional Analysis: the Spatial Dimension of Development Policy*, Westview Press, Boulder and London, 1985.

Scheidegger, Adrian E.: *Morphotectonics*, Springer, Berlin, 2004.

Sebba, J.: *Geography for All*, London, David Fulton, 1996.

Sokhi, B.S. and S.M. Rashid : *Remote Sensing of Urban Environment*, Manak, Delhi, 1999.

Steven, A. : *Grand Canyon: Monument to Catastrophe*, Institute for Creation Research, London,1994.

Vergne, Jack: *Hard Rock Miner's Handbook*. Tempe/North Bay: McIntosh Engineering, 2003.

Viles, Heather: *Biogeomorphology*, Basil Blackwell, Oxford, 1988.

Wainwright, John; Mulligan, M.: *Environmental Modelling: Finding Simplicity in Complexity*, John Wiley and Sons Ltd, London, 2003.

Walter Leal: *Innovation, Education, and Communication for Sustainable Development*, New York, Peter Lang Publishing, 2006.

Wendy J. Gordon : *The Economics of Copyright, Developments in Research and Analysis* Edward Elgar, 2003.

Williams, Paul W.: *Catena Supplement 25*, Catena Verlag, Cremlingen-Destedt, Germany, 1993.

Yang, Xin-She: *Introductory Mathematics for Earth Scientists*, Dunedin Academic, Edinburgh, 2009.

Index

❑❑❑